A GOLD HUNTER'S GUIDEBOOK
to
ARIZONA'S PLACER GOLD

Author panning for gold on Hassayampa Creek

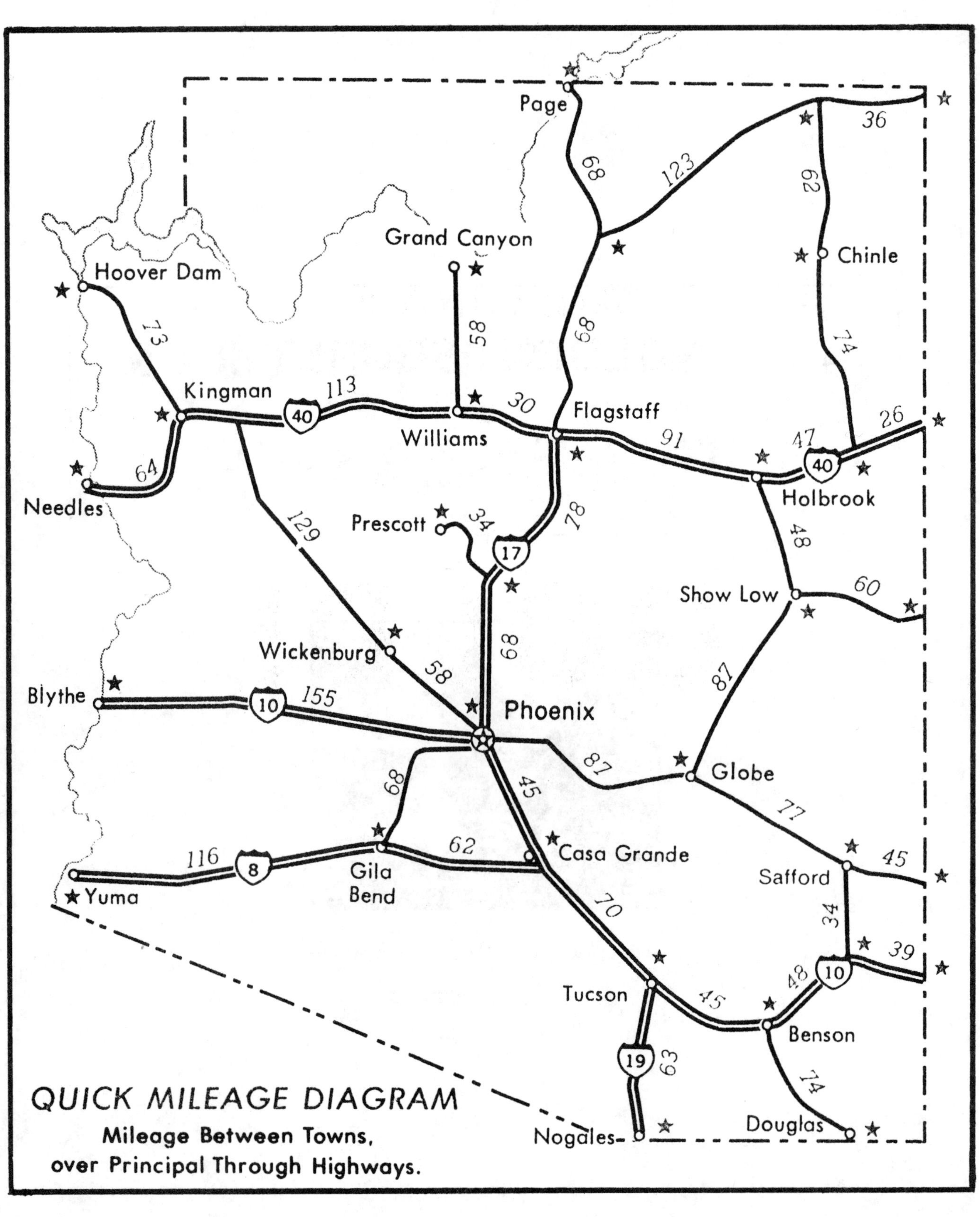

QUICK MILEAGE DIAGRAM
Mileage Between Towns, over Principal Through Highways.

ARIZONA'S GOLDEN SECRET

OR

HOW TO GET YOUR SHARE
OF
DESERT

GOLD!

BY
RONALD S. WIELGUS

DEDICATION

I dedicate this book to Alyn J. Wielgus, my wife. She accompanied me on many of my earlier trips and shared in the joys, and the disappointments, in my quest for the elusive metal. Although unable to make the trips to the gold fields now due to a physical disability, she does accompany me in spirit and with encouragement.

ISBN 0-9635601-0-7
ARIZONA'S GOLDEN SECRET
OR HOW TO GET YOUR SHARE OF DESERT GOLD!
Second Edition (Revised).

Cover Photo: Gold nuggets recovered by the author, Gravel Gertie Claim, Ash Canyon, Huachuca Mountains, Cochise County, Arizona; photo by the author.

CONTENTS

ACKNOWLEDGMENT

I thank my son, Roger, for the many hours he spent reviewing portions of the manuscript. He offered many helpful suggestions incorporated into this book, and provided several photographs of us together mining the gold.

Las Guijas Placers - 1992. Upper: On the road to the Las Guijas Mountains in the distance; Lower: Roger unloading drywashing equipment at an upper wash area, Las Guijas Mountains.

ABOUT THE AUTHOR

Ronald S. Wielgus is a professional architect and has lived in Arizona since 1959. He has been prospecting for Arizona's gold for 33 years ever since the thrill of finding his first nugget in a dry wash northwest of Black Canyon City. From that time on he has hiked into some of the more remote gold-bearing washes of the Bradshaw Mountains and suction-dredged many of its gold-bearing creeks including Lynx Creek before it was closed to all suction dredges and motorized equipment. On several of these dredging ventures he enlisted the aid of his son, Roger, who learned how to look for and get the gold. He owned and mined placer claims in the Huachuca Mountains for almost 10 years, and now mines Arizona's rich, desert placers with drywashers of his own design. Along with his interest in gold is a fascination with forgotten mining camps and ghost towns and he has metal-detected such places as Palmerlee, Alexandra, McCabe, Bueno and a host of others.

Roger Wielgus, the author's son, is a native born Arizonan and computer programmer/analyst who is now living in southern California. He first became interested in gold prospecting while still in grade school and accompanied his father on many trips to the gold-bearing creeks of the Bradshaw Mountains where together they operated a 2-1/2 inch suction dredge. Subsequently, he acquired his own 3 inch suction dredge which he used on their claims in the Huachuca Mountains. He has extensive knowledge on prospecting and finding gold in stream beds and is a proficient gold-panner. In addition, he has a keen interest in Arizona ghost towns and also has metal-detected obscure sites such as McCabe, Bueno and Gillette as well as cabin sites of long-ago prospectors. His career demands have not dampened his interest in prospecting, and he looks for the gold whenever time allows.

OVER, Figs. A - F. A: Huachuca Mountains as seen from Hwy. 92, looking Northwest - 1957; B: Lynx Creek below Lynx Lake - 1960; C: Road to Hassayampa Creek from Wolf Creek Campground - 1960; D: Las Guijas Mountains as seen from Las Guijas Road, looking Southwest - 1992; E: Santa Rita Mountains as seen from Melendrez Pass, Melendrez Pass Road west of Greaterville - 1992; F: Author testing Ophir (Hughes) Gulch one mile west of Greaterville - 1992.

In the early 1980's, a prospector made a major gold discovery in Arizona in an area not previously identified as a mineral district. The prospector who discovered the Copperstone deposit noticed a few very small, isolated outcrops of mineralized rock and staked a claim. The opening of the Cyprus Copperstone gold mine in La Paz County in 1987 is proof that prospectors have not found all the gold deposits in Arizona. This mine will produce 50,000 to 60,000 troy ounces of gold per year during its projected life span of 5 to 6 years and is expected to double Arizona's gold production.

***

This old prospector of a century ago won his gold with a cradle, or rocker, a device still used today.

PREFACE

Man has had a love affair with gold long before recorded history. What was this strange fascination with an object that could neither clothe nor feed him? In the pre-dawn of our history survival was paramount, yet he cherished the yellow metal although it had no utilitarian value.

In this world of impermanence it seemed that gold was eternal. It did not tarnish, corrode or decay and always remained bright, shiny and unchanging. It acquired deep mystical qualities, eternal life itself in an age when the human life span measured only a few decades, and it was rare.

About 10,000 years ago man domesticated grain. This allowed him to cease his nomadic wanderings in the constant search for food and with this new found freedom the seeds of civilization sprouted. Afterward come the domestication of cattle and man was well on his way toward the destiny we all share today.

Later, as civilization took root and grew a ruling hierarchy came into being. Gold became the symbolic connection between the ruling god king on earth and his eternal continuation in the afterlife. Ordinary persons could not own this royal metal; this noble gold reserved for the king.

About 5,500 years ago certain civilizations in the Middle East became prosperous. They had surpluses of grain and cattle but shortages of other things they desired or needed. To acquire those things they did not have they bartered and traded what they had. A class of merchants and traders sprang into being. It soon became apparent that barter was cumbersome. For some societies that had gold, and silver, but limited surplus resources barter was not practicable. In those prosperous societies that had surplus resources however, gold became a symbol of wealth. What better way was there to show success than by personal adornment with that rare metal that conveyed beauty, symbolism and mystique. Although known and admired for its beauty too, silver did not possess those qualities so unique to gold. Foremost, it tarnishes and corrodes upon exposure to the elements.

Eventually a medium of exchange developed between those that had surpluses of products but desiring others, and those that had few surpluses or none at all but having gold and silver. Because silver is not so rare as gold it also became a medium of exchange, but because of the very fact that it isn't as rare as gold it isn't as valuable. Therefore, it took much more silver to conduct transactions than gold, and it tarnishes and corrodes! These are hardly the "qualities" one wants in exchange for valuable commodities. The one metal possessing all the qualities necessary for use as a medium of exchange is gold. It became *the* measure of value of goods and services. It became money and coveted for the importance and power it bestowed upon those in possession of it. For with gold, rulers could raise armies to make war on their neighbors, bribe others to kill for them, pay off more powerful ones and do a host of other deeds both good and bad.

This recognition of the power of gold drove men to great lengths to obtain it. History is replete with the unimaginable cruelty and misery men have inflicted on themselves and others in their intoxicated desire for this elusive metal. To possess it they have endured hardships that stagger the imagination.

The world of earlier times was much different then than it is now. You were either born into nobility, the merchant class, serfdom or slavery and if the last, virtually no way to get out of it. Locked in from birth to death, a serf was always a serf, a slave always a slave. If you could obtain gold however, then you could buy your way out. You could gain title, importance, power and respect. This, then, was the real meaning of wealth conveyed by the possession of gold.

However, the limited opportunities prevailed to get gold. The discovery of the New World in 1492, and all of its riches, opened the doors of opportunity to the hundreds of thousands seeking escape from the prison of their miserable existence. It began with the Conquistadors and culminated with the Forty-niners. After one last gasp in the gold rushes of the Klondike in the Yukon, and at Nome, Alaska, gold slowly faded from the scene as a way of escaping one's fate, or did it?

The world was changing. It became the modern one we now take for granted. Opportunities expanded. There were other ways to work oneself out of poverty. Gold seemed to have lost its importance and president Franklin Delano Roosevelt issued his executive order that called in all the gold held by the citizens of our country. Eventually the U. S., and all other industrialized countries the world over, went off the gold standard. Paper currency, and finally, clad coins with no intrinsic value became the money of the day.

In spite of all this, gold has endured as the true standard by which we measure value. It has been so for thousands of years. It has endured when all other forms of money have failed because it has intrinsic value, value alone. It doesn't need a declaration proclaiming that it is money. Gold *is* money!

At this very moment deep in the jungles of the Amazon and Papua New Guinea, a new rush is on as thousands of impoverished souls seek a way out of grinding poverty in their mad pursuit of gold. That bright, shiny, elusive yellow metal so coveted over centuries has not lost its allure nor forsaken its promise.

***

11

INTRODUCTION

There is a saying that, "Silence is golden." When it comes to gold the silence is almost deafening. Why is that?

First is the high intrinsic value of the metal. Gold is *real* money. When it comes to money, very few people want to talk about it with strangers. You will have a real problem, therefore, in obtaining firsthand information about prospecting for gold in Arizona. There seems to be an unwritten code of silence among the fraternity of gold miners when it comes to discussions about gold. To get any useful information out of them on where you can go to get the gold is almost futile. If you can get them to tell you anything, it will be purposely vague, misleading or inaccurate to steer you away from *their* domain and *their* gold. After all, why should they share with you the secrets of where and how to get the gold that they, themselves, are after?

Second is the paucity of detailed written information on where to look for and how to get the gold, especially regarding Arizona. There are many books about gold but very few have sound advice on where to look, equipment to use and the techniques necessary to successfully finding and winning the gold.

This book reveals to the amateur those areas and methods he needs to know to become a successful gold hunter and in the process relax the mind and rejuvenate the body. In this book I will concentrate only on placer gold, Arizona placer gold. This is gold that has eroded from the veins and stringers in the mother rock and has washed down into the gullies, arroyos, washes and streams as flakes and nuggets. It is "free gold" because it is now free from the grip of the host rock. It is also free for the taking. Lode gold is gold still encased in the host rock. It requires more sophisticated and expensive methods of mining and is not so free for the taking.

To find gold in Arizona we must go to known gold areas. Arizona has many gold-bearing areas, some discovered hundreds of years ago by the Spanish, worked by the Jesuit fathers with the labors of the native peoples, then abandoned and forgotten. Mexican and Anglo-American prospectors, in the middle to late 1800's, rediscovered the old Spanish diggings and discovered new ones on their own. Continually harassed by the Apache these miners reaped a rich harvest in gold, but **they didn't get it all.**

Prospectors have walked over virtually every square foot of Arizona at least once in their search for gold and other minerals. As a result, they have defined most of the placer gold fields. However, there is always the chance of making a significant gold discovery, particularly in the drier, less-visited desert areas of the state. We have at our disposal equipment far more sophisticated for finding and recovering gold than those early prospectors who relied more on their eyesight and intuition than anything else.

Although the early miners exhausted the richest placers, there are many areas of the state hardly touched since the depression years of the 1930's. These areas still yield good gold and the mountains continue to feed flakes and nuggets to the washes and streams, year after year. Arizona has much gold but you have to know where to look for it.

The quest for Arizona's gold continues unabated. From the large mining companies to the most humble individual, the search goes on in the deserts and mountains of our state.

It seems that most of Arizona's newcomers and visitors also want to go out and look for gold. Many become disappointed and discouraged. They usually quit in frustration. Others are more persistent and may get rewarded by a few colors and flakes. However, each has gotten out to experience the extraordinary beauty of Arizona, and whether successful or not, in the process engaged in healthful exercise.

I hope that this book will provide the stimulus for everyone interested in finding gold to go out, enjoy this great state, and be successful in unlocking the door to Arizona's golden secret.

***

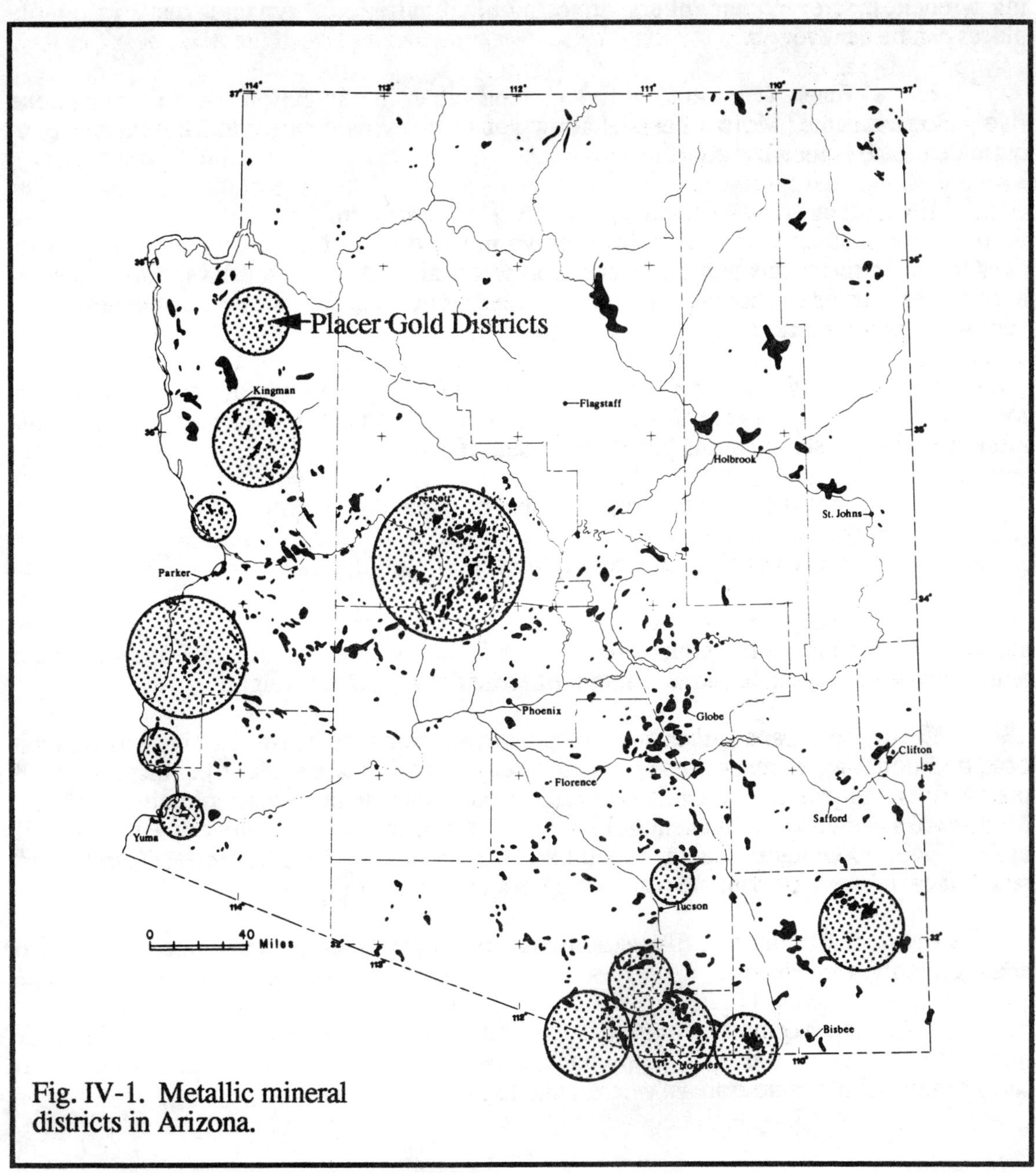

Fig. IV-1. Metallic mineral districts in Arizona.

CHAPTER I

EQUIPMENT

To prospect for and mine placer gold you need, at a minimum, the following equipment.

1. A reliable vehicle, preferably high centered (4x4 is desirable but not necessary), with *good* tires and at least one *good* spare. Along with this, tools for repair including wrenches, screwdrivers, pliers, annealed black wire, etc. Arizona roads in remote places can be very rough.

2. Water. Carry at least five gallons, either in one-gallon containers or in one five-gallon container. More water will allow you to pan your concentrate, but the basic requirement here is for survival.

3. Food (It's hardly equipment yet necessary for survival). Packaged dried food is good insurance in case of breakdown in the desert. Note that water has priority over food. In the desert you can survive on water alone for several days, but food soon becomes exhausted or spoiled. Then, you are in big trouble. Without water the desert is a very unforgiving place.

4. Gasoline. Carry a full tank and at least two extra gallons in one-gallon metal containers as insurance. Many areas are remote and far from services, but this minimum requirement should allow you to reach help if needed.

5. Shovel or shovels, long-handled or D-handle, or both.

6. Hatchet (This is particularly useful for cutting brush).

7. Small digging tools such as a stainless steel trowel; pick; pry bar; large screwdriver used to work cracks and crevices in the bedrock; metal spoon; stiff bristle whisk broom, and a small sledge hammer of around three pound weight.

8. Gold pans, either metal or plastic. I prefer the old-style, high-sided metal ones traditional with prospectors of bye gone days, but this is a matter of choice. A small plastic finishing pan is useful in working down concentrate but not absolutely necessary. Many books geared to the amateur gold hunter describe panning techniques (see Chapter II on WHERE TO GO and Chapter IV on OPERATING YOUR EQUIPMENT), so I won't repeat them here. Skill with the pan only comes with practice.

9. Plastic four or five-gallon buckets for gravel and concentrate, at least two and preferably four or more.

10. Ore bag or sack made of heavy cotton duck in which to carry samples. For many years I used an old U. S. Army backpack that I still have, and use occasionally, to carry samples from remote areas, where I had to hike.

11.   Screens, in at least two sizes, 1/4 and 1/2 inch mesh, the circular kind that will fit over the mouth of the plastic bucket. These depend on the size of nuggets expected. You will have to screen and classify the gold bearing gravel for further processing.

12.   Small container or vial for the gold. Clear plastic film containers work best as they are unbreakable and have a tight-fitting cover. Glass vials are subject to breakage and loss of your gold, so you must be very careful with these.

13.   Tweezers (to pick up the smaller flakes). Always pick up your gold with a container under it large enough to catch it if dropped. A gold pan works well. Never pick up your gold so that it can drop into the gravel of the desert; you probably won't find it again.

14.   Insect repellent. This is necessary unless you have a natural immunity to the biting effects of rapacious flies prevalent in the desert in April-May and again in August-September; get the kind that contains at least 40% DEET and use liberally.

15.   Small first aid kit.

16.   Snakebite kit. Arizona has many different kinds of rattlesnakes. Most avoid direct, hot sunlight and usually hibernate during the cold winter months. They can be quite active on overcast days during the rainy season and especially at night during the warmer time of the year. I have come across rattlesnakes many times in the Arizona desert, spying them coiled under mesquite bushes, under or around large boulders, in dense brush and at puddles of water left from summer thunderstorms. Give them a wide berth. Please don't kill them as they are necessary to the ecological balance. Usually they will move out on their own, but if they don't you always can find another spot to work. The desert is a big place with room for everyone.

17.   Flashlight (with fresh batteries and spares).

18.   Matches, kitchen type strike-anywhere, sealed in protective plastic bags.

19.   The gold hunter may want to add several more items at his discretion such as toilet paper, paper towelling, personal hygiene items, etc.

OPTIONAL EQUIPMENT FOR THE SERIOUS PROSPECTOR:

1.   **Drywasher** (Figs. 5 & 6). This one piece of equipment is necessary if you want to work the desert placers . Drywashers are readily available in manually operated or motorized versions from certain manufacturers of gold recovery equipment and from individuals scattered throughout the state. These are not cheap. If you can, buy a used drywasher through your local gold equipment dealer, or better yet, from an individual disposing of his machine and in need of cash. I custom-make my machines tailored to my specific needs. These work as well, if not better, than the manufactured ones. To work at *top* efficiency drywashers must be fed gold-bearing gravel that is absolutely dry, and classified, that is, screened down to no larger than 1/2 inch mesh and preferably smaller as close sizing improves the recovery. Watch out for nuggets too large to pass through the screen. Feed the machine at a constant, slow rate with the riffle tray inclined and full of material, and all the riffles covered evenly and puffed not too strongly by the bellows or blower. A drywasher, *properly designed and operated*, will recover most of the gold present in the

gravel. You will find the gold behind the riffles, that is, on the uphill side of the riffles and sitting in the dead-air spot along with much black sand.

2. **Portable Screen** (Fig. 7). I designed and built this piece of equipment to facilitate speedy screening of placer gravel when using a drywasher. The screen, mounted on legs, is self adjusting. Screen material is one-half inch square hardware cloth; however, you may use smaller mesh. A chute under the screen channels screened materials directly into a plastic bucket. In operation, I simply toss a shovel full of gravel onto the screen and gravity does the rest.

3. **Suction Dredge** (Fig. 8). For many years I operated a 2-1/2 inch suction dredge on many of the small, gold-bearing, live streams and creeks in Arizona. Even the small flow present in intermittent creeks during the rainy seasons enabled me to use a suction dredge of this size. On my former claims, after the snow melted and the creek began to flow, I used the 2-1/2 inch dredge and my son used a 3 inch one. Anything larger than 3 inch is, in my opinion, impracticable on Arizona's streams as many do not possess sufficient flow to support the operation of larger suction dredges. The availability of water, generally from winter rains or from snow melt in the higher elevations, controls suction-dredging. Most creeks go dry in May. Some replenish after summer rains, and may flow in September and October, then go dry again. Suction-dredging is an iffy proposition in a state short on water but can be extremely productive under the right conditions. If you have sufficient water available, this is the way to go as you will be able to process far more gravel than with any drywasher.

4. **Gold Wheel** (Fig. 9). This hand-fed, self-panning motorized device separates the fine gold and smaller flakes from the black sand concentrate with a gentle spray of water. It's a necessity when handling large amounts of concentrate and recovery is very efficient. There are several manufacturers on the market and the wheels are costly items, but they make up for that in fast, thorough and efficient cleanup of concentrate. However, if expense is a factor then you must resort to the time-honored method of panning which will take up much time and may result in loss of fines. Panning successfully is dependent on the skill of the panner and this only comes with practice.

5. **Metal Detector** (Fig. 10). I use two VLF metal detectors in the field. The design of one machine is only for finding gold nuggets, and the other is an all-around, very high quality detector that automatically tunes itself to ground conditions. I do not use discrimination in their operation. Although you can find nuggets in the washes and small arroyos through patient operation of a metal detector, I use mine mainly to check the screenings for any nuggets too large to pass through the sieve being used. Sometimes I check the shoveled overburden just to be on the safe side. Metal detectors can be quite expensive, particularly the better makes. A quality machine can make up for this in its ability to detect gold, and gold is what we're after.

There are other devices available for the serious gold hunter with ample financial resources such as, jigs, tables, concentrating bowls, etc. I prefer to go after the gold for the pure enjoyment of it and to be moderately successful in the endeavor. So, it is not necessary to spend thousands of dollars on equipment that you may use very little. Unless, you are in for it on a large scale, a commercial venture geared to making a profit...which is not the purpose of this book.

***

CHAPTER II

WHERE TO GO

Arizona is one of the most highly mineralized states in the continental United States and noted primarily for its copper deposits, but not too much has been published about its gold deposits. The Arizona Bureau of Mines did two, major works many years ago. One documents lode deposits; the other describes the state's placer deposits. The author considers the Arizona Bureau of Mines Bulletin Number 168 entitled *GOLD PLACERS AND PLACERING IN ARIZONA*, originally written in 1933 and republished many times since, to be the bible for all serious gold hunters in this state. This book is still available from the University of Arizona Press in Tucson, and from book stores and dealers in prospecting equipment for a small price, or may be found in the larger public libraries in the state. It features a section describing gold panning techniques.

Another work, entitled *PLACER GOLD DEPOSITS OF ARIZONA* by Maureen G. Johnson, essentially repeats the information in Bulletin Number 168 with a little more historical information added. It also is available from book stores and prospecting equipment retailers and state libraries.

A series of articles written by Edgar B. Heylmun, Ph.D., featuring placer localities in Arizona has appeared from time to time in the *CALIFORNIA MINING JOURNAL*. Each article is an in-depth study of a specific locality, with a short history, geological description, excellent reference map and a personal assessment of the area. These articles are an invaluable source of information on some of the lesser known areas. The serious gold hunter might want to contact the publisher at P. O. Box 2260, 9032 Soquel Drive, Aptos, CA 95001, for further information on obtaining these articles.

The U. S. Geological Survey (USGS) publishes topographic maps that are useful in orienting yourself in the field. They also publish technical mineral and geological information that may be helpful. For more detailed information on USGS, contact them at USGS, Branch of Distribution, P. O. Box 25286, Federal Center, Denver, CO 80225. Other good sources of technical information in Arizona are the Arizona Geological Survey, University of Arizona, Tucson, Arizona 85721, and the Department of Mineral Resources, 1502 W. Washington, Phoenix, Arizona 85007. Ask for their current price list of publications.

Because Arizona is a desert state that is dry most of the time, you should gear your efforts toward the dry washes and the use of a drywasher. Here are a few localities where the novice can find gold.

THE SOUTHEAST CORNER OF ARIZONA (Fig. 1):

A highly mineralized region centers on the **Las Guijas Mountains** southwest of Tucson near **Arivaca**. Virtually all the washes issuing out of this range carry gold. There are many claims and private holdings on the south side of the mountains, fewer on the north side. Many of the claims are invalid from lack of assessment work. There are some state lands here as well. It is a very remote and rugged area with few roads. You can reach it driving south from Tucson.   Take Interstate I-19 to Arivaca Junction near Amado, then Arivaca Road and turn right onto a dirt road marked by a mailbox with the numerals "46" on it. Follow this road west, but take the left fork. Do not turn onto the road marked "Arroyo Seco Ranch." Continue west approximately 5 miles or until you cross the cattle guard west of a few ranch buildings and windmill. Before you reach these landmarks you will drive through a flat area that has an abandoned tank truck off to the side. The road here becomes a quagmire after heavy rainfall and only negotiable by a four-wheel drive vehicle. Do not turn off at any of the side roads until you are beyond the last cattle guard. You are now entering the gold-placer area. There are a few dirt roads heading toward the mountains that are negotiable in a passenger car, but be careful. Some of the side branches are impassable except in a high centered pickup truck or four-wheel drive vehicle.

I have drywashed in this area in the heat of the pre-monsoon months and found it tolerable. The months of May and June, although very hot, are excellent in that the gravel is bone dry and the area devoid of people. When cooler fall weather arrives it can be a very pleasant place but then one sees more placer miners. If the dry weather holds, then mining can continue into the winter months. Contrary to the report in Bulletin Number 168, I found drywashing to be excellent.

All the washes in this area carry gold. It is necessary to get close to the mountains to access the bedrock and the larger gold. Bedrock in the upper reaches of the washes is very close to the surface, hardly more than a foot below, but it gets progressively deeper as you go downstream. In many places the bedrock exposes on both sides of the washes. This gives you and indication of what it looks like as it goes under the stream gravel. Some bedrock, heavily cracked and fissured, usually extends across the wash (see Fig. 3C). These cracks and fissures are of great interest to the gold hunter. The gold becomes trapped in them. They usually don't extend much deeper than a few inches into the bedrock. A pick or a sharp-pointed shovel is all you'll need to rip up this bedrock to get at the gold. The gold here is bright, mostly flat and coarse, ranging from flakes to nuggets exceeding 3/16 inch. The chances are very good for finding large nuggets in these upper washes. Much of the gold I found still has iron oxide attached to it indicating a local origin.

Farther out from Arivaca are the placer fields around **Oro Blanco**. Much of this locale is under claim and worked heavily by miners for placer and lode gold. There are many open shafts in the area and wandering about can be dangerous. Almost all the washes in this area carry whitish, bright, fine gold and flakes. The area can get much rain in the summer rendering drywashing impracticable.

The **San Luis Mountains** are southwest of **Arivaca** and are very difficult to reach. Here one needs a vehicle designed for rough roads; a passenger car is not practicable. This is a good area to use a drywasher as it is a very dry place except for the occasional summer thunderstorm. The gold is coarse but seems limited to those washes below the lode mines. However, you cannot rule out other washes in the more remote parts of this district.

Closer to Tucson are the **Santa Rita Mountains**. The specific area is **Greaterville**, which was one of the premier placer localities of the state in the late 1800's. The early miners recovered many large nuggets from the washes to the west and south of Greaterville and large nuggets still may lie undiscovered awaiting the lucky prospector. To reach this area from Tucson, drive southeast on I-10 to the turnoff at State Highway 83.

Go south on 83 to Greaterville Road, then west on Greaterville Road to Melendrez Pass Road. Head south on Melendrez Pass Road to Greaterville. When you reach Greaterville there will be several roads branching out to the West, South and East. Take the west or south roads, but keep within a mile and a half of Greaterville if you take the south road, or you will drive out of the placer area.

Almost all the washes heading west into the mountains contain placer gold. Again, the closer you get to the mountains the less the overburden on bedrock. Try to set up on one of the smaller tributaries as far from the main roads as possible. Gold hunters have heavily worked this area but it still has much coarse gold. Drywashing here is only possible in the dry months of May-June and September-October because these densely wooded mountains receive a high rainfall during the summer monsoon and some snow in the winter. Creek flow is unreliable or insufficient if your mining and recovery equipment are dependent on local availability of water.

The **Huachuca Mountains**, where my family and I owned and mined claims, is a little known range that has good placer gold. The specific locality is **Ash Canyon**, reached by State Highway 92 south from **Sierra Vista**. A fire devastated much of this area in the late 1980's. In fact, I was placer mining on my Sparkle Plenty Claim the very morning that the fire roared over the ridge. The burn took out thousands of acres of vegetation. Subsequent heavy rains on the denuded slopes resulted in landslides and several additional feet of overburden deposited in the creek. Many of the access roads washed out in the process. Before this it was relatively easy to get to Ash Creek and its gold.

Take the dirt road west marked "Ash Canyon" beyond the cattle guard. Upon crossing the cattle guard you enter the Coronado National Forest, away from the private holdings and into the gold area. The major wash on the left side is Ash Creek and it carries gold throughout its length. On the right side is an unnamed wash that carries fine gold. When you come to the fork in the road a sign posted there reads "Lutz Canyon," that goes right, and "Ash Canyon," to the left. The Lutz Canyon road dead ends at the Miller Peak Wilderness Area less than a mile beyond. The Ash Canyon road dead ends about a mile beyond the fork. Taking the Ash Canyon road you cross Lutz Creek at a hairpin bend in the road. This was the site of the author's Gravel Gertie Claim sold in 1990. Upstream from the crossing, and not more than 50 yards in, my son and I took out very coarse, heavy gold nuggets from bedrock gravel. This gold, of a deep color, contained much quartz still attached and indicated a very close source. Although I searched for the source of the gold over many years using VLF metal detectors, deep-seeking detectors and the classical method of test-holes, I never did locate it. The overburden here ranges from a foot to well over four feet thick and contains many boulders. Most of the gold was close to or on the bedrock but some was within a foot of the creek's surface. In 1982, I suction-dredged a one-pennyweight nugget that was not more than a foot below the surface!

The most likely spot for very large nuggets is just before the creek reaches the road. The road crosses over the creek bed on a deep layer of fill. This fill, acting as a dam, has stopped farther migration of gold downstream and it also is here that the flow of water has slowed due to a widening out of the creek. The amount of overburden at this spot is several feet. However, there is a real possibility of finding a rich concentration of nuggets, some weighing over an ounce, that are sitting here on the bedrock.

The south fork of Ash Creek, farther on up the road, attracted the attention of many because of its very coarse gold. The source of the gold was probably the Yaqui Lode Mine upstream. A nugget found nearby in 1911 weighed 32 ounces! Sporadic mining by individuals still continues as evidenced by the small holes dug in the creek bed.

Those concerned about valid claims might want to consider the wilderness area where the road dead ends in Lutz Canyon. A short walk will take you into the wilderness area and access to Lutz Creek, which is the same one that passed through our Gravel Gertie claim. You may not use motorized equipment or locate a mining claim in the wilderness area. If you plan on doing any prospecting in the wilderness area, check first with the Coronado National Forest District Ranger in Sierra Vista for the latest regulations.

In wet years both forks may have a good spring flow that should enable you to use a suction dredge. The flow may last a few months or be as short as a few weeks. It is difficult to predict when, and if, the creeks will be flowing. In some years they don't flow at all. Try the months of February through April. The dry months are May, June, October and November. Because the elevation is over 5000 feet, snow is possible in the winter months and the ground may freeze.

The **Dos Cabezas Mountains**, southeast of **Wilcox**, also have coarse gold. These mountains are far from civilization and a long drive from Tucson by way of Insterstate I-10 to Wilcox, then southeast to **Dos Cabezas**. Roads in the mountainous areas are rough and a passenger car not recommended.

Almost all the washes heading into the mountains from the little community of Dos Cabezas carry gold. It is necessary to get into the upper reaches of the smaller washes to avoid much overburden. In the dry seasons this is good drywashing country, but the summer rains render the gravel too moist to process through a drywasher. Unfortunately, the rainfall is not enough to develop a sufficient flow of water in the washes to enable the use of a suction dredge.

Much of the area is under claim but there could be many invalid claims resulting from the failure of claimants to perform their annual assessment work. A look through Cochise County records at the Recorder's office in Bisbee should show which claims have lapsed and what areas are open (see Chapter VI on CLAIMS AND CLAIMING).

THE NORTHWEST CORNER OF ARIZONA:

For a review of this remote corner of the state, besides Bulletin Number 168, there is a very good book entitled *THE ART OF DRYWASHING OR FINDING GOLD IN THE DESERT* by Otto E. Lynch that is available in most bookstores.

THE BRADSHAW MOUNTAINS, PRESCOTT AND VICINITY (Fig. 2):

This is one of the most highly mineralized areas of the state concerning gold. Almost all creeks and streams originating in these mountains carry gold that is a bright, brassy yellow in color. I have suction-dredged the **Hassayampa, Lynx, Turkey** and **Big Bug** creeks. All of these carry gold. In addition, I have panned gold from **Soap** and **Slate Creeks** northwest of Black Canyon City. By far, the richest creek worked by my son and I has been Lynx, but now closed to suction dredges and motorized equipment of any kind. All of these creeks carry sizable nuggets. There is a real possibility of finding nuggets over an ounce in weight. However, don't overlook the tremendous amount of fine gold that is present that really adds up. Save all of your black sand concentrate and use a gold wheel for separation, or run these through a tumbler with a little mercury for amalgamation. It may surprise you to see how much gold you will recover with these methods. See Chapter V on how to separate the gold from your concentrate.

A good place for the novice is Lynx Creek about 7 miles southeast of Prescott just upstream from **Lynx Lake**. This area is open to all, but only shovels and pans permitted. You usually find "color" in your pan and needn't worry about being on someone else's claim. The Forest Service has set aside this site as a recreational gold panning area, withdrawing it from mineral location and mining claims. The gold that I recovered with my suction dredge was very bright and flat, and many flakes were larger than match heads. The fine gold that I separated from the concentrate equalled or exceeded in weight that which I recovered as flakes. My son panned this creek upstream to the community of **Walker** and found gold in every pan. Some of the largest placer gold nuggets ever found in Arizona came from Walker.

Hassayampa and Big Bug Creeks are easy to reach on good dirt roads. You should have no difficulty in a passenger vehicle. Each creek is capable of yielding nuggets ranging in size from that of match head and bigger. Suction-dredgers find nuggets exceeding one-half ounce in weight almost every year. Gold occurs throughout the length of both creeks.

Hassayampa Creek is accessible by driving south of **Prescott** on the Senator Highway.   When you pass through the community of **Groom Creek,** an area of summer homes, turn right onto the dirt road signed "Wolf Creek Campground" and drive to the lower campground turnoff. This is a U. S. Forest Service campground with picnic tables. A short dirt road leads south from the campground down to the creek that flows throughout the year.

On both sides of the creek, upstream and down, there is evidence of past and present placer mining. In some places, craters made by the activities of gold hunters cover the hillsides. Accompanied by my wife, I found the gold here to be bright yellow and the largest pieces about the size of a match head although some others took out much larger nuggets. My best returns came from working behind and under the boulders in the creek. See ABOUT THE AUTHOR and the photograph showing me suction-dredging Hassyampa Creek in 1981.

Big Bug Creek is northwest of **Mayer.**   It is accessible from Poland Junction just west of the Prescott-Phoenix highway. With my wife assisting, I operated a sluice box many years ago on one of the tributaries and found very small gold that was no bigger than a pinhead, although the potential is there for much larger gold. I had to dig into the gravel bars to get at the paydirt that I shoveled into the sluice. Diverting the flow of the tributary through the sluice provided the water necessary to wash the gold-bearing gravel over the riffles.

There is much evidence of past and present mining activity. Big Bug Creek itself is a popular gold dredging creek with many of today's gold hunters and commercially mined in the depression years of the 1930's. Water is reliable in the spring from mountain snow melt.

To get to Turkey Creek go west out of **Mayer** on the dirt road leading into the mountains. As you get into the mountains you will see flowing streams and creeks. Some of these may host small parties of miners using suction dredges. Turkey Creek itself is at the road forks. There is a cabin here shown on maps as **Goodwin.** The south road goes to Crown King; the north road to Prescott. This is high elevation country with Ponderosa Pine. It is cool in summer and can be snowbound in winter. The best times for suction-dredging are late April through June and again after the summer rains. Much of this creek was under claim when worked by my son and I years ago and open ground was very limited. This situation may have changed considerably over the years and may probably change again due to proposed legislation in the mining law.

WEST OF PHOENIX:

The **Weaver Mountains** and **Rich Hill,** near the ghost towns of **Octave** and **Stanton** east of **Congress,** are rich in gold. Rich Hill itself was notable for potato-sized nuggets found under rocks and in the small washes at its very summit! To reach this locality, drive west out of Phoenix to Wickenburg and go north on U. S. Highway 89 past Congress approximately 5 miles. Go right on the dirt road to the ghost town of Stanton. The town is no more than a few intact dilapidated buildings with tin roofs surrounded by a wire fence and no trespassing signs. A couple of miles east of here was the town of Octave. If you continue on the dirt road past Octave you will loop back to Wickenburg but I recommend returning by the same way you came which is shorter and not as rough.

My son and I, many years ago, suction-dredged Weaver and Antelope Creeks that drain south on either side of Rich Hill, and found much fine gold hardly bigger than a pin-head. This gold was a bright, brassy yellow. Bedrock is very close to the surface in many of the smaller washes and all the small pockets in it had gold. You may have to recycle your water to dredge Weaver Creek and the water will muddy up very badly making visibility at your suction nozzle virtually nil. We had to go by feel alone using our bare hands to follow the paydirt. It was very dirty work but worth it.

This place can be very hot in the summer and there is very little shade. We confined our dredging activity to the very early spring months of February through April when it is cooler and the creeks are flowing. Be very careful when near the rocky places as rattlesnakes abound and you may see several on the dirt road warming themselves in the rays of the early morning sun.

Much of this area is under claim and some of the people living there are not friendly to strangers. Check out land status records before you do any mining here.

The **Vulture Mountains** and vicinity are a good drywashing area but much of it now is under claim. I prospected several dry washes in this area in the 1970's with some success, but accidentally wandered on to Vulture Mine property and told to leave. A thorough search of the records in the Maricopa County Recorder's Office and the BLM offices in Phoenix is necessary to determine the status of claims in this area. It's proximity to Phoenix and Wickenburg makes it a choice spot to visit, though.

To reach this area drive west from **Wickenburg** on U. S. Highway 60 approximately 8 miles and drive south approximately 15 miles on the gravel road on the left signed "Vulture Mine." The rugged mountains on the left are the Vulture Mountains and Vulture Peak is the prominent rock spire. To the West and South are the drywashing gold fields. The washes in the low mountains to the West carry fine gold. The gravel in them becomes thinner as one gets higher in elevation and it is necessary to hike into the area to reach them. Summers are hot and the best time is any other time. Winter rains may prevent drywashing the gravel that must be dry.

Driving farther south from here you will come to the famous Vulture Mine that Henry Wickenburg discovered in 1863. You may take a walking tour from 9:00 a.m. to 5:00 p.m. every day of the week from September 15 to May 11. During the summer the mine is open from Thursdays through Sundays. There is an admission charge. You may inquire at the Chamber of Commerce in Wickenburg about the current price before making the drive out or buy your admission at the mine. There are picnic areas available for day use. For those who want to get gold the easy way the mine encourages visitors to pan its (salted?) gold bearing gravel at the sluice box.

San Domingo Wash and tributaries north of **Morristown** and the Wickenburg Highway is a good area for drywashing even though miners worked the ground over many times and untouched spots are now hard to find. The smaller tributaries will be your best bet. The area's proximity to Phoenix makes it an easy drive and a good one-day visit. The time to prospect there is from the months of October through April when the desert is cooler. Winter rains in December and January will usually put down any drywashing. The extreme northern part of the district is a wilderness area and motorized equipment not allowed.

Much farther west is **Quartzsite**, a classic placer gold locality. Snowbirds from all over the U. S. converge on the area with metal detectors and one of the largest nuggets found in the U. S., with a metal detector, was from this area. Much of the gold found here is virtually on the surface and those using metal detectors have raked the desert clean in many areas in their fever to get at it. To get the drywashing gold it is necessary to reach the washes issuing out of the mountains west of Quartzsite. Several dirt roads from Quartzsite will take you there. Vehicular ways crisscross the entire area around Quartzsite and you may camp just about anywhere in the desert. This is a very pleasant place to visit in the winter months when temperatures are mild and you won't be lacking company.

Finally, just before reaching **Ehrenberg** west of Quartzsite, Interstate I-10 passes through the mountains and one of the better placer fields, the **La Paz**. Almost all the washes heading into the west side of the mountains north of the Interstate carry gold, but the best washes are off limits as they occur on the Indian Reservation. Again, try to get to the smaller tributaries closer to the mountains to do your prospecting. Many washes dissect the terrain making desert driving a little difficult. This is another hot place in the summer but can be a very enjoyable one in the winter.

THE SOUTHWEST CORNER OF ARIZONA:

This part of the state with **Yuma** as its headquarters holds much promise. Unfortunately, a great part of this region is off limits due to military, wildlife refuge and Indian reservation withdrawals. This is the pioneering area of the state for the serious gold hunter. A drywasher is a necessity to test the gravel of the many isolated mountain ranges. As one of the more remote parts of Arizona, it presents challenges to man and vehicle and can be a very dangerous place should your vehicle breakdown. I recommend that persons going into the isolated ranges do so with a buddy following in a second vehicle. If you must go it alone, let someone you know where you are going, the duration of your stay, and when you expect to return. Because of the nature of the terrain, isolation and a harsh summer climate only the most experienced gold hunters need apply.

The most accessible of the areas and least hostile is the **Gila City Placers** on **Monitor Gulch** east of Yuma in the **Gila Mountains**. The area's proximity to Yuma makes it a popular spot with the winter visitors trying their luck in the smaller washes. To reach the placer field from Yuma drive east on U. S. Highway 95. When the road turns north it will parallel the Southern Pacific Railroad tracks and take you around the northwest side of the mountains. Just before you reach the Gila River turn right onto the dirt road that follows alongside the tracks. Take this road and drive about three miles to the placer area.

The prominent wash draining north out of the mountains is Monitor Gulch and its gold is deep under much overburden. However, other washes carry gold just west of this gulch. Again, you should direct your prospecting efforts toward reaching bedrock in the smaller washes and tributaries, and to the spurs where tributaries come together (see Fig. 4A). Most of the gold in these washes is small, about pinhead size. If you are lucky, you could get rewarded with a nugget. Heylmun, in the July 1988 issue of the *California Mining Journal*, Volume 57, Number 11, gives a good account of this area and includes a map.

THE NORTHEAST CORNER OF ARIZONA:

No significant gold deposits, either placer or lode, occur on the Colorado Plateau. The Mogollon Rim escarpment defines the southern limit of this region.

A WORD OF CAUTION:

I cannot overemphasize that most old mine shafts and workings are extremely dangerous and may be on private land or already under a mining claim. If you go into an old mining district or ghost town to hunt for artifacts or to prospect, you may be trespassing on a mining claim or private land and taking private property. Also, any object over 100 years old on Federal land is an antiquity protected under Federal law. The law does not specifically prohibit you from using a metal detector to search for artifacts, but you should exercise caution. For a review of claiming, see Chapter VI on CLAIMS AND CLAIMING.

***

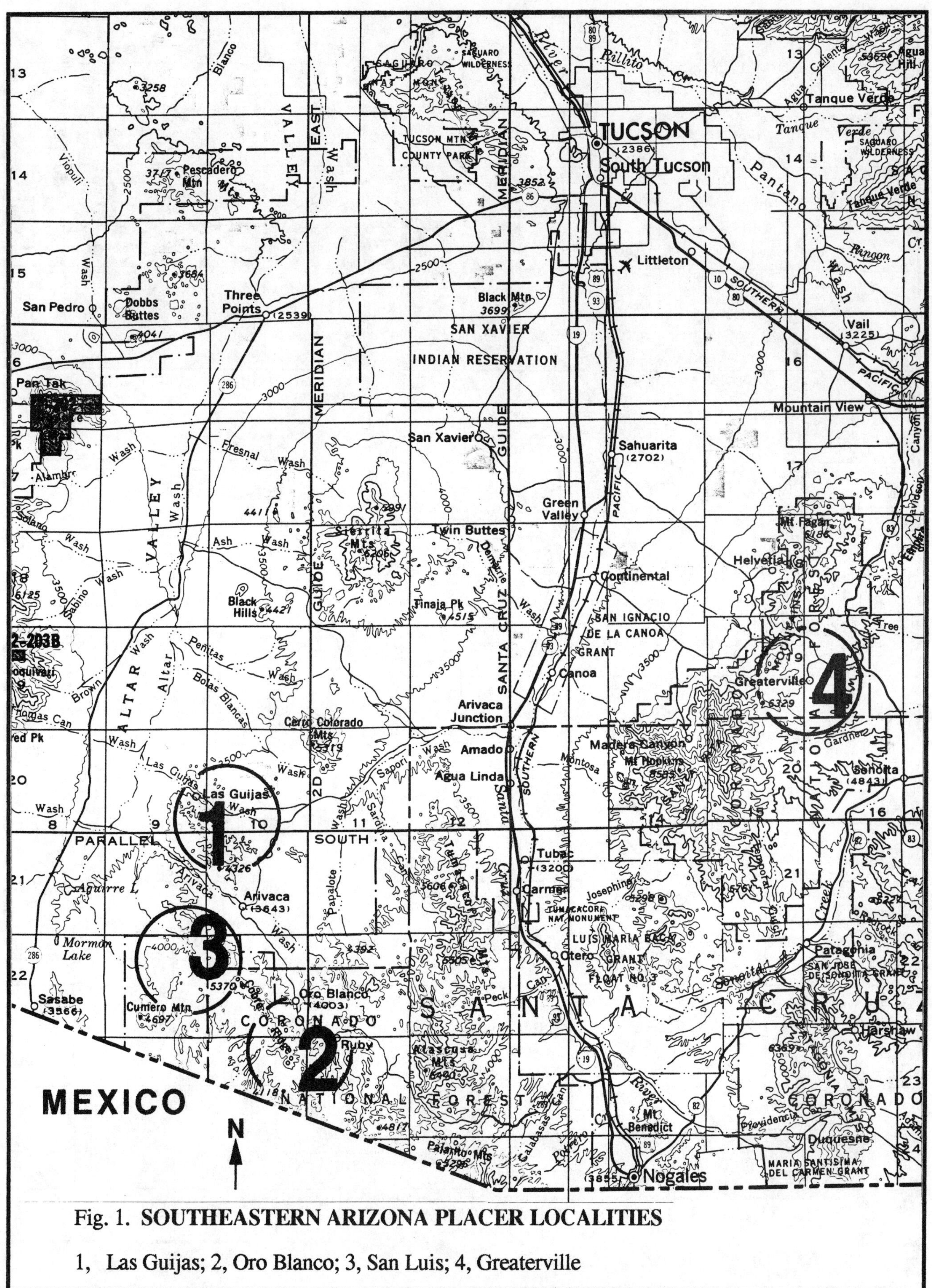

Fig. 1. SOUTHEASTERN ARIZONA PLACER LOCALITIES

1, Las Guijas; 2, Oro Blanco; 3, San Luis; 4, Greaterville

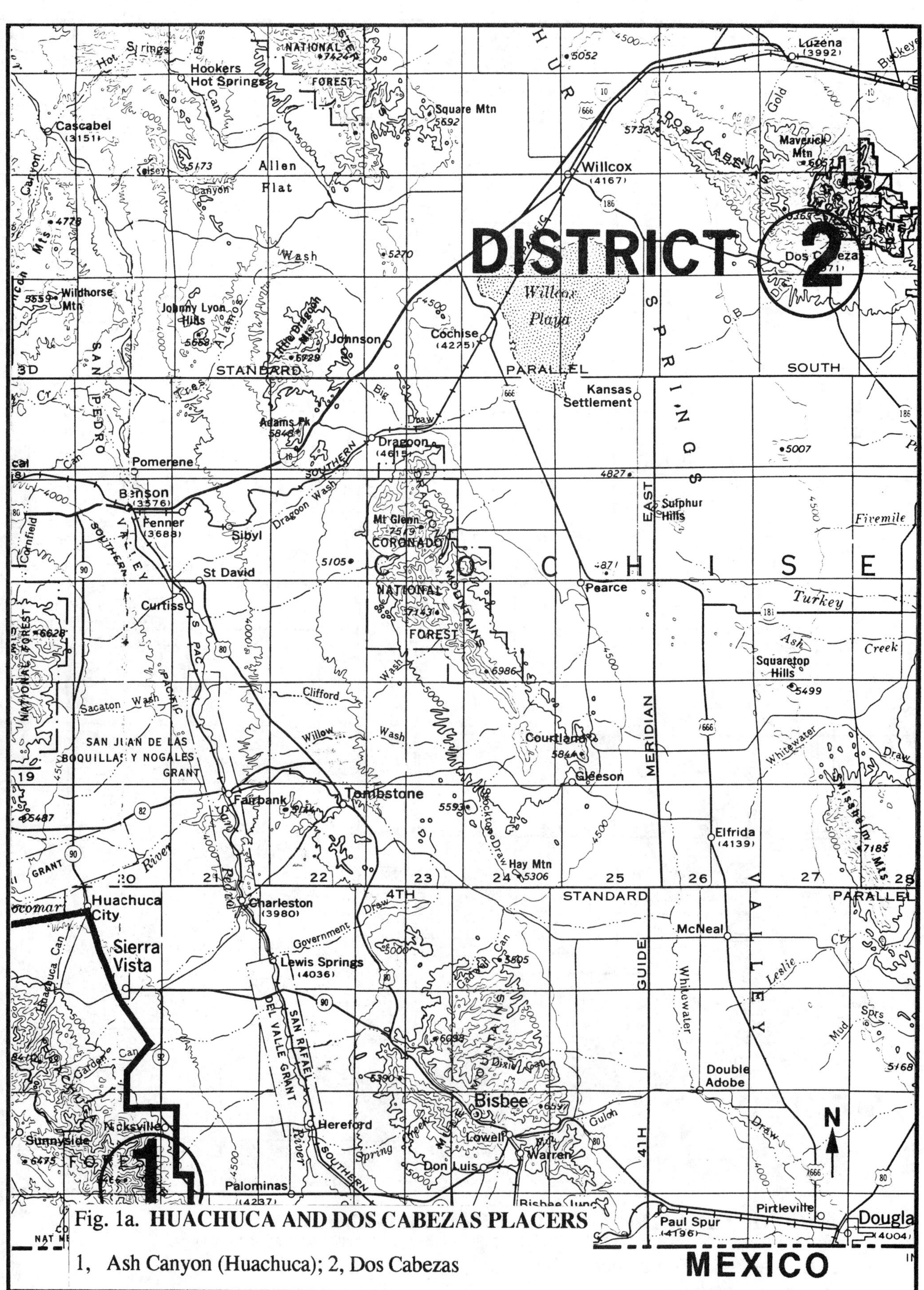

Fig. 1a. **HUACHUCA AND DOS CABEZAS PLACERS**

1, Ash Canyon (Huachuca); 2, Dos Cabezas

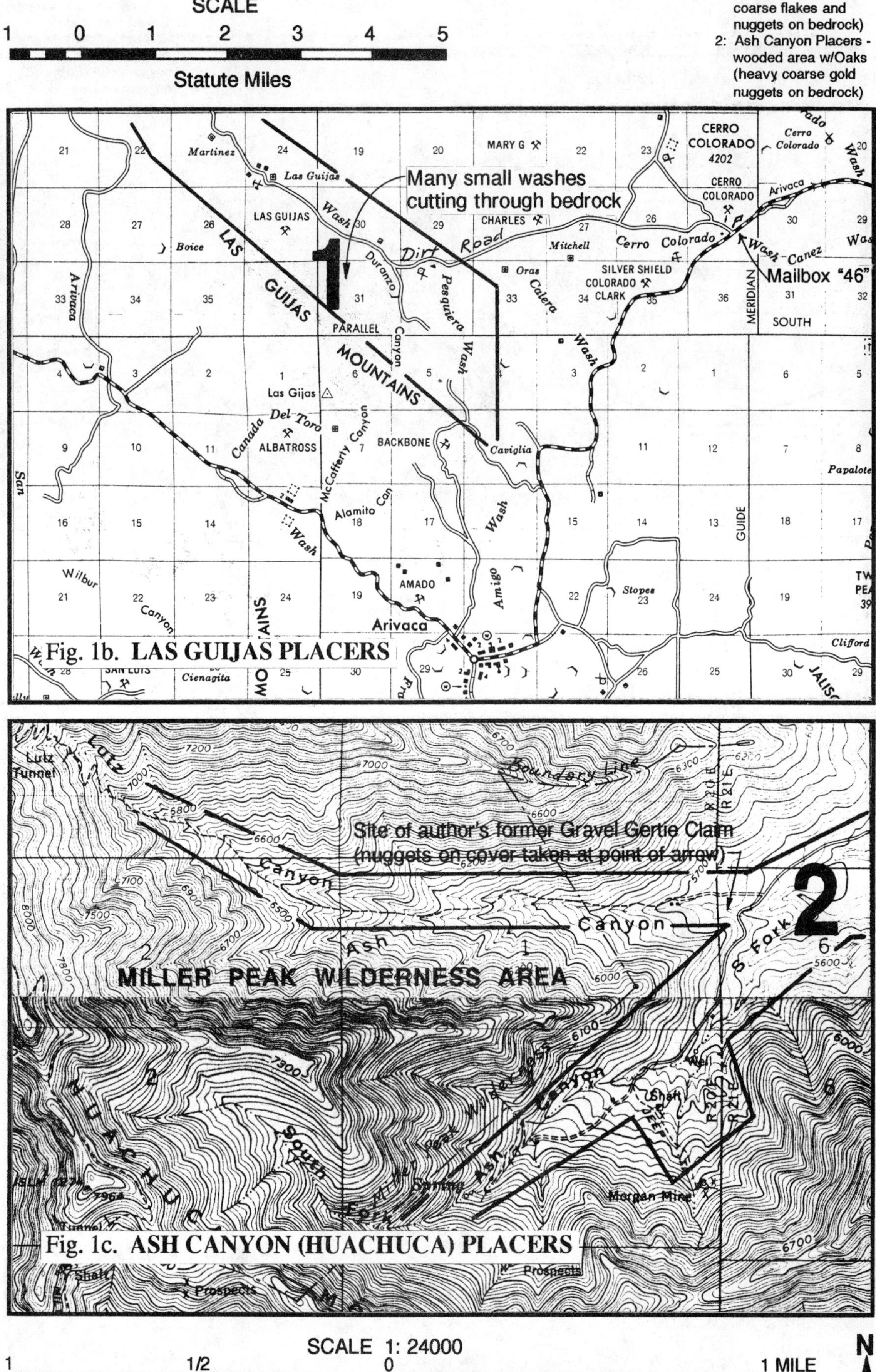

SCALE
1 0 1 2 3 4 5
Statute Miles
1: Las Guijas Placers - rough dirt road (bright, coarse flakes and nuggets on bedrock)
2: Ash Canyon Placers - wooded area w/Oaks (heavy coarse gold nuggets on bedrock)
Many small washes cutting through bedrock
Mailbox "46"
CERRO COLORADO 4202
CERRO COLORADO
Cerro Colorado
Arivaca
MARY G
CHARLES
Mitchell
Cerro Colorado
SILVER SHIELD COLORADO CLARK
Oras
Calera
Martinez
Las Guijas
LAS GUIJAS
Boice
Arivaca
LAS
GUIJAS
MOUNTAINS
PARALLEL
Wash
Duranzo
Pesquiera Canyon
Wash
Dirt Road
MERIDIAN
SOUTH
Wash-Canez
GUIDE
Las Gijas
Canada Del Toro
ALBATROSS
McCafferty Canyon
BACKBONE
Caviglia
Wash
Amigo Wash
Papalote
San
Wash
Alamito Can
AMADO
Arivaca
Stopes
TW PEA 39
Clifford
Wilbur
Canyon
MOUNTAINS
SAN LUIS
Cienagita
JALISC
Fig. 1b. LAS GUIJAS PLACERS
Lutz Tunnel
Lutz
Boundary Line
Site of author's former Gravel Gertie Claim (nuggets on cover taken at point of arrow)
Canyon
Ash
Canyon
Canyon
S Fork
MILLER PEAK WILDERNESS AREA
HUACHUCA
South Fork
Shaft
Morgan Mine
Tunnel
Shaft
Prospects
Prospects
Fig. 1c. ASH CANYON (HUACHUCA) PLACERS
SCALE 1: 24000
1 1/2 0 1 MILE
N

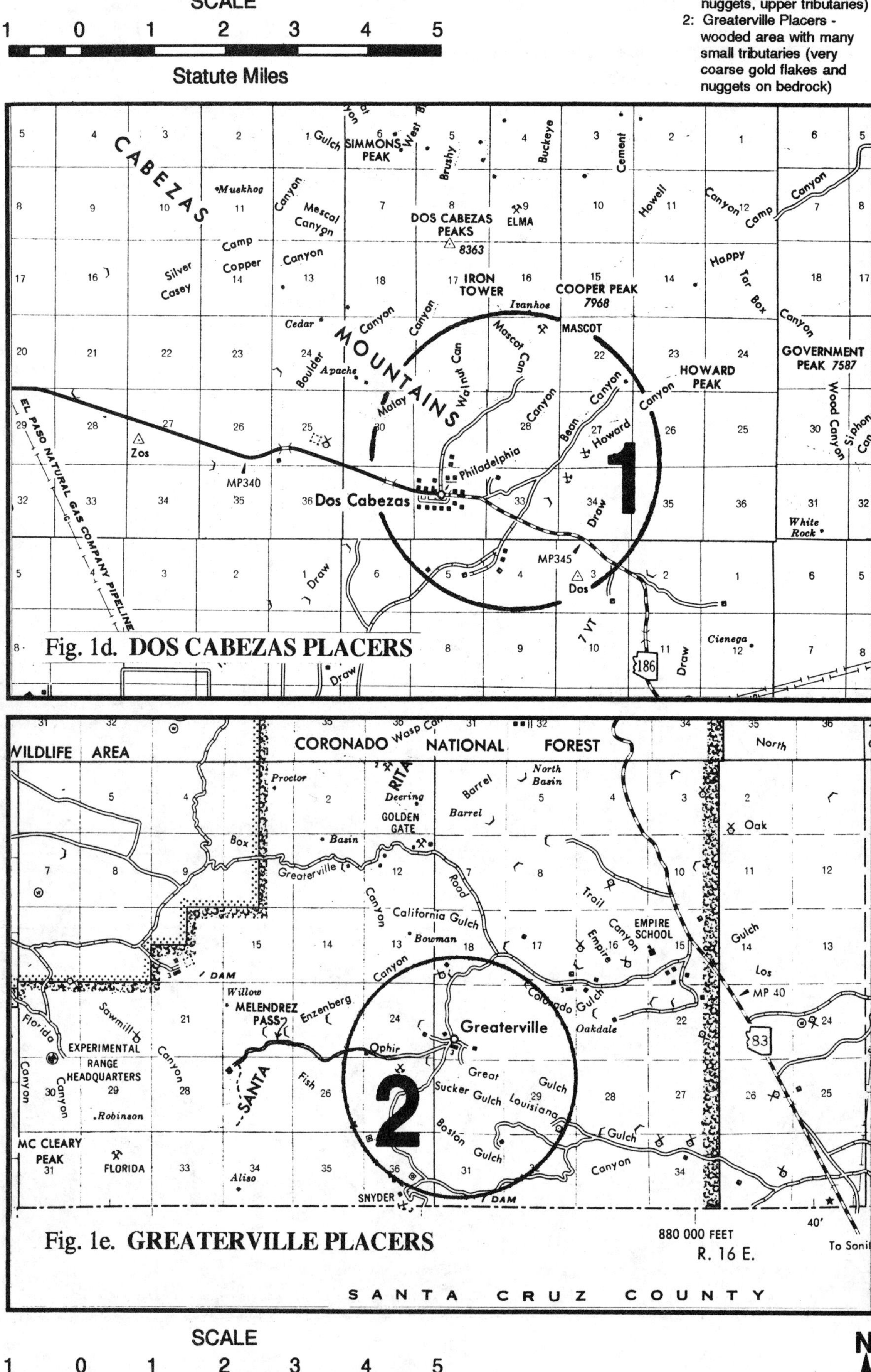

Fig. 1d. **DOS CABEZAS PLACERS**

Fig. 1e. **GREATERVILLE PLACERS**

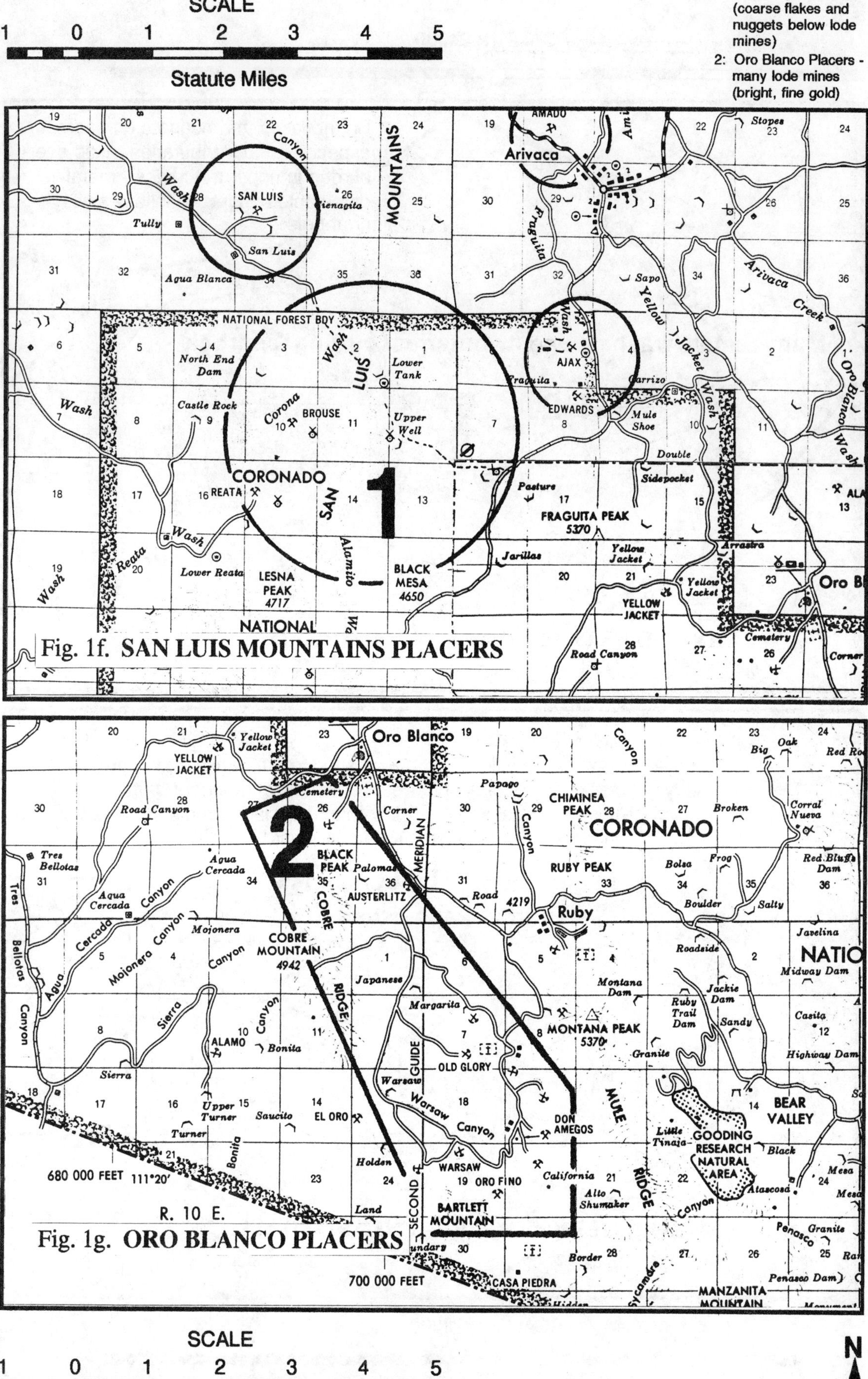

Fig. 1f. **SAN LUIS MOUNTAINS PLACERS**

Fig. 1g. **ORO BLANCO PLACERS**

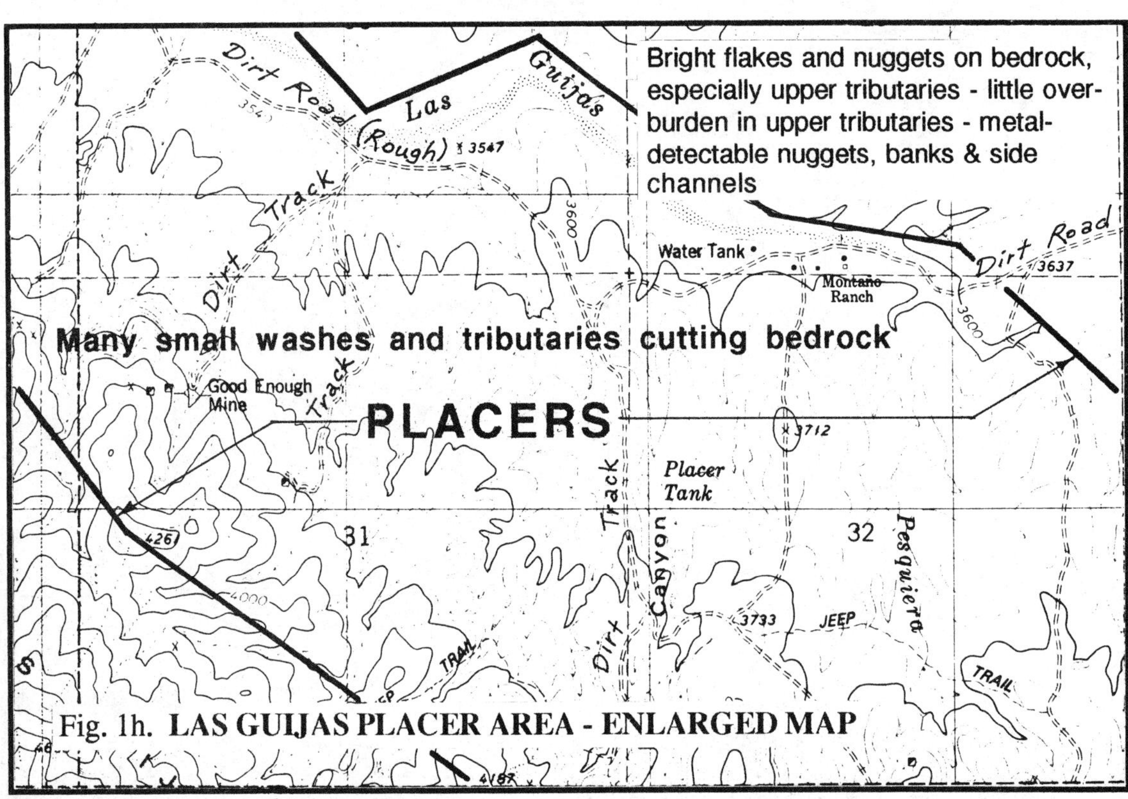

Fig. 1h. LAS GUIJAS PLACER AREA - ENLARGED MAP

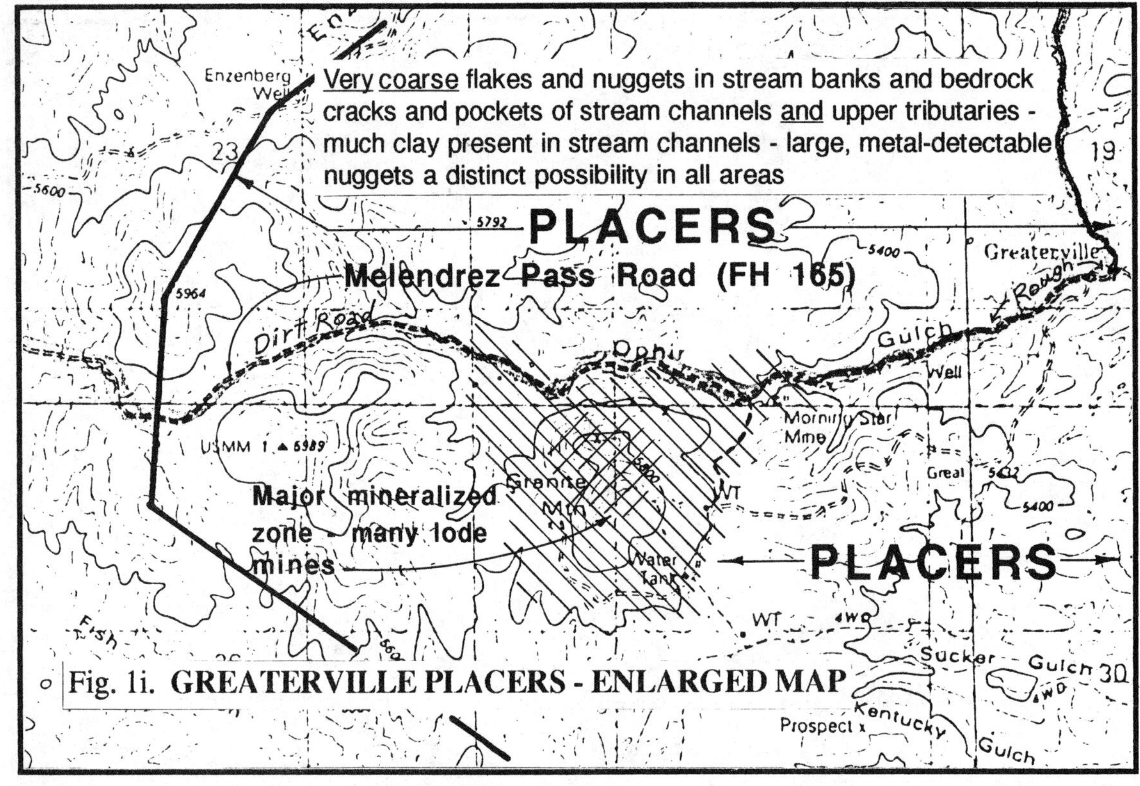

Fig. 1i. GREATERVILLE PLACERS - ENLARGED MAP

N

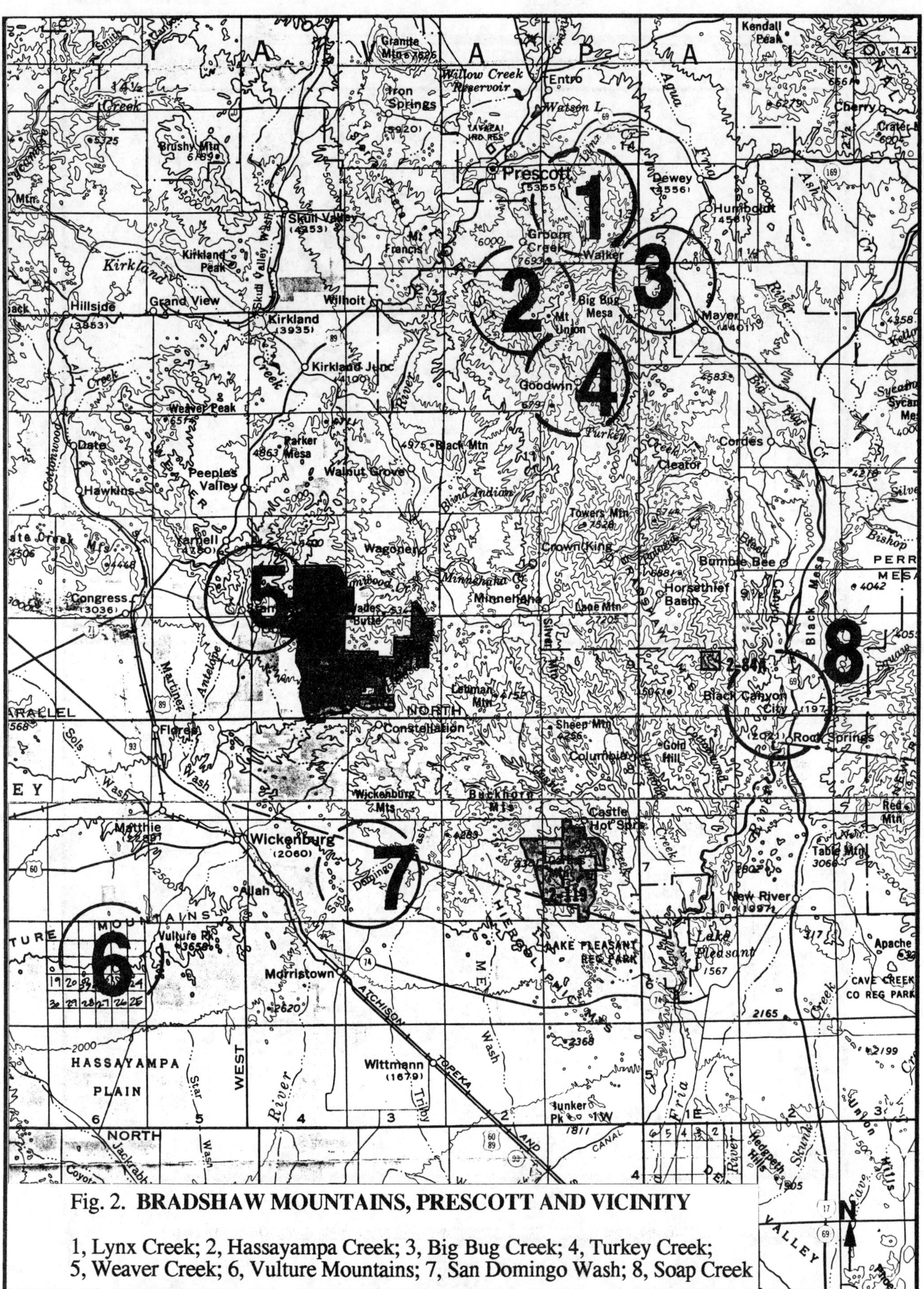

Fig. 2. BRADSHAW MOUNTAINS, PRESCOTT AND VICINITY

1, Lynx Creek; 2, Hassayampa Creek; 3, Big Bug Creek; 4, Turkey Creek;
5, Weaver Creek; 6, Vulture Mountains; 7, San Domingo Wash; 8, Soap Creek

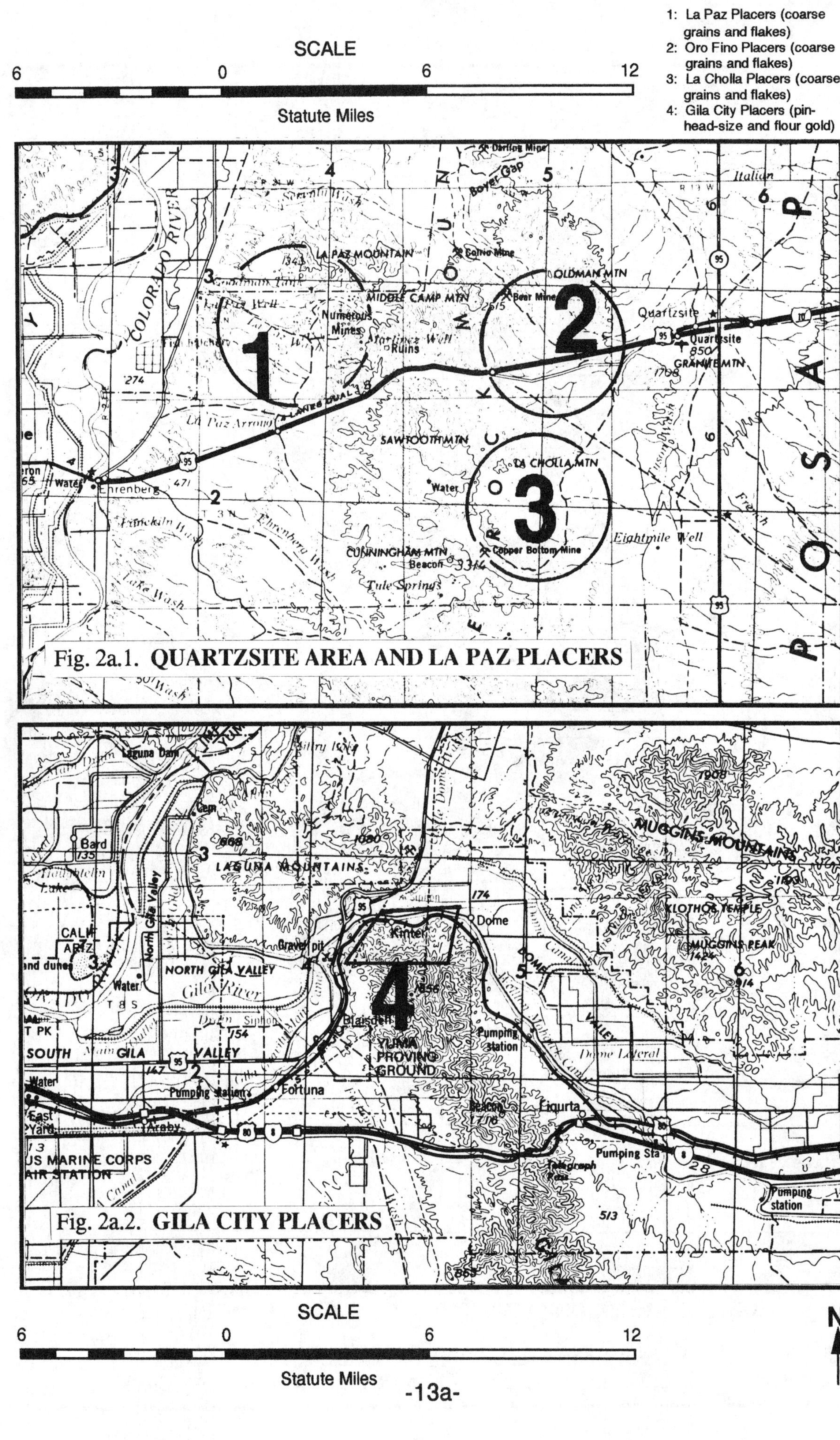

SCALE
6   0   6   12
Statute Miles
1: La Paz Placers (coarse grains and flakes)
2: Oro Fino Placers (coarse grains and flakes)
3: La Cholla Placers (coarse grains and flakes)
4: Gila City Placers (pinhead-size and flour gold)
COLORADO RIVER
LA PAZ MOUNTAIN
MIDDLE CAMP MTN
OLDMAN MTN
Quartzsite
GRANITE MTN
Numerous Mines
Martinez Well
Ruins
SAWTOOTH MTN
LA CHOLLA MTN
Water
Ehrenberg
CUNNINGHAM MTN
Beacon
Tule Springs
Eightmile Well
Copper Bottom Mine
Darling Mine
Boyer Gap
Italian
La Paz Arroyo
Fig. 2a.1. QUARTZSITE AREA AND LA PAZ PLACERS
Laguna Dam
MUGGINS MOUNTAINS
LAGUNA MOUNTAINS
KLOTHO TEMPLE
MUGGINS PEAK
Bard
CAL. ARIZ.
NORTH GILA VALLEY
Gila River
Gravel pit
Kinter
Dome
YUMA PROVING GROUND
Blaisdell
SOUTH GILA VALLEY
Fortuna
Ligurta
Pumping station
Araby
US MARINE CORPS AIR STATION
Pumping Sta
Pumping station
Fig. 2a.2. GILA CITY PLACERS
SCALE
6   0   6   12
Statute Miles
N

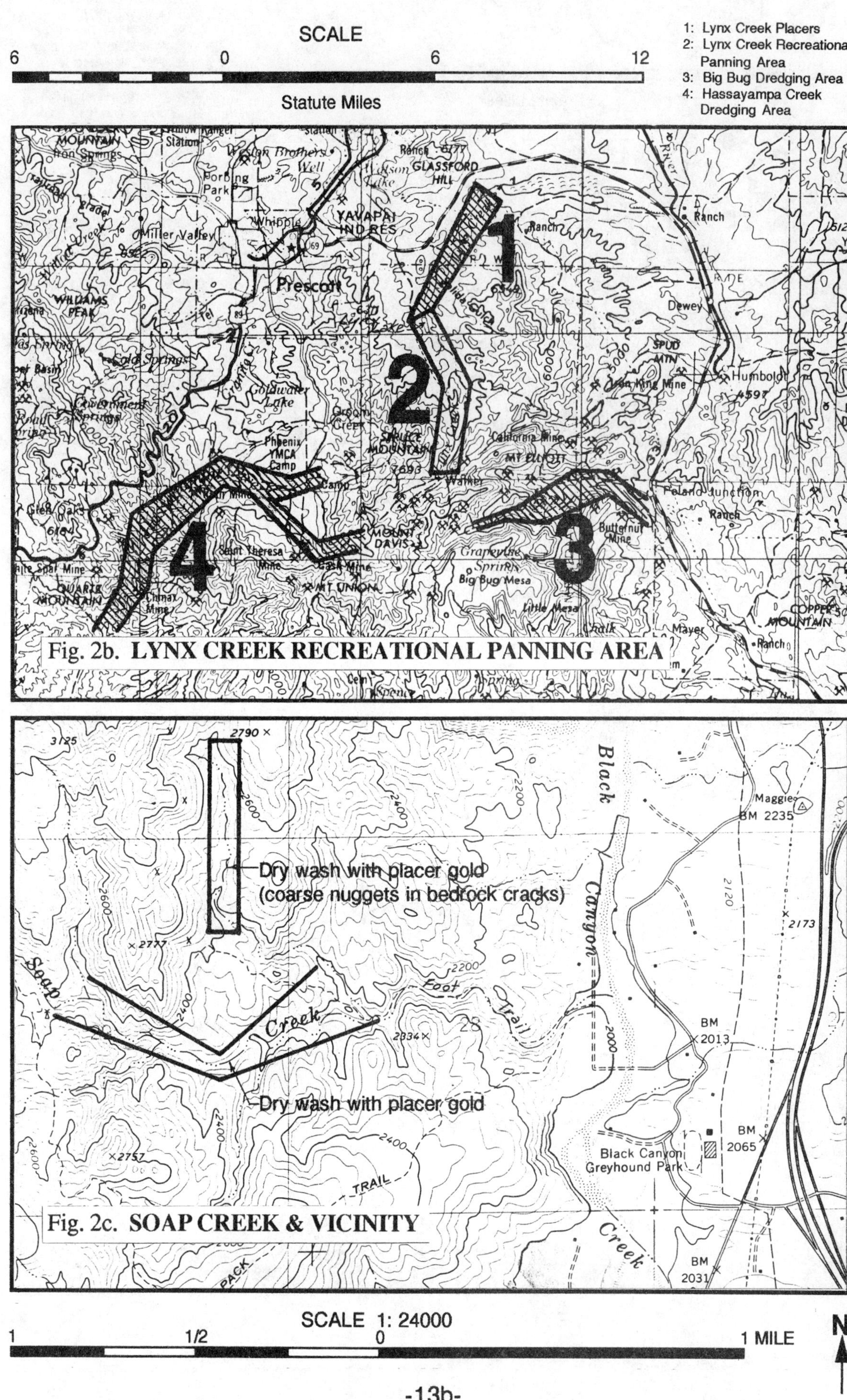

SCALE
6   0   6   12
Statute Miles

1: Lynx Creek Placers
2: Lynx Creek Recreational Panning Area
3: Big Bug Dredging Area
4: Hassayampa Creek Dredging Area

Fig. 2b. LYNX CREEK RECREATIONAL PANNING AREA

Dry wash with placer gold
(coarse nuggets in bedrock cracks)
Dry wash with placer gold
Fig. 2c. SOAP CREEK & VICINITY

SCALE 1: 24000
1   1/2   0   1 MILE
N

SCALE

6   0   6   12

Statute Miles

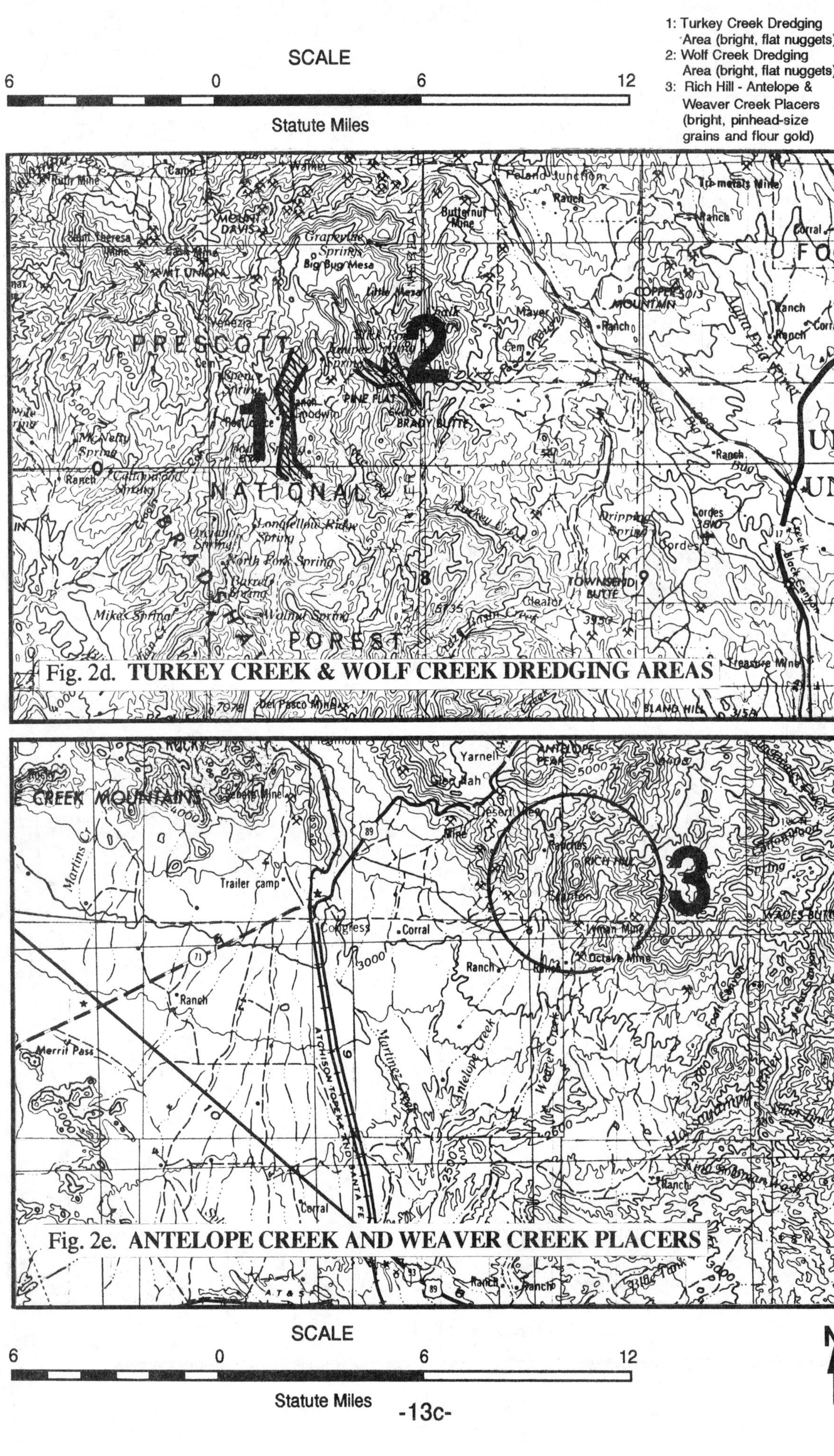

Fig. 2d. **TURKEY CREEK & WOLF CREEK DREDGING AREAS**

Fig. 2e. **ANTELOPE CREEK AND WEAVER CREEK PLACERS**

SCALE

6   0   6   12

Statute Miles

N

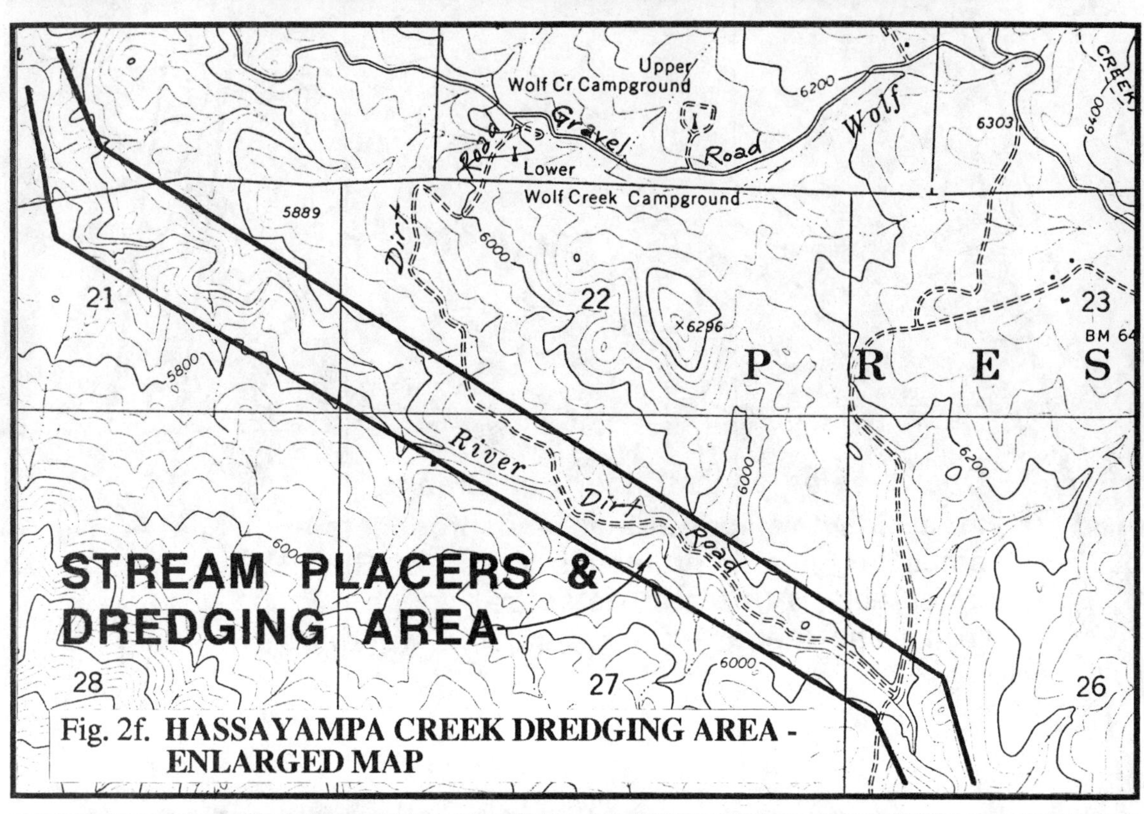

Fig. 2f. HASSAYAMPA CREEK DREDGING AREA -
ENLARGED MAP

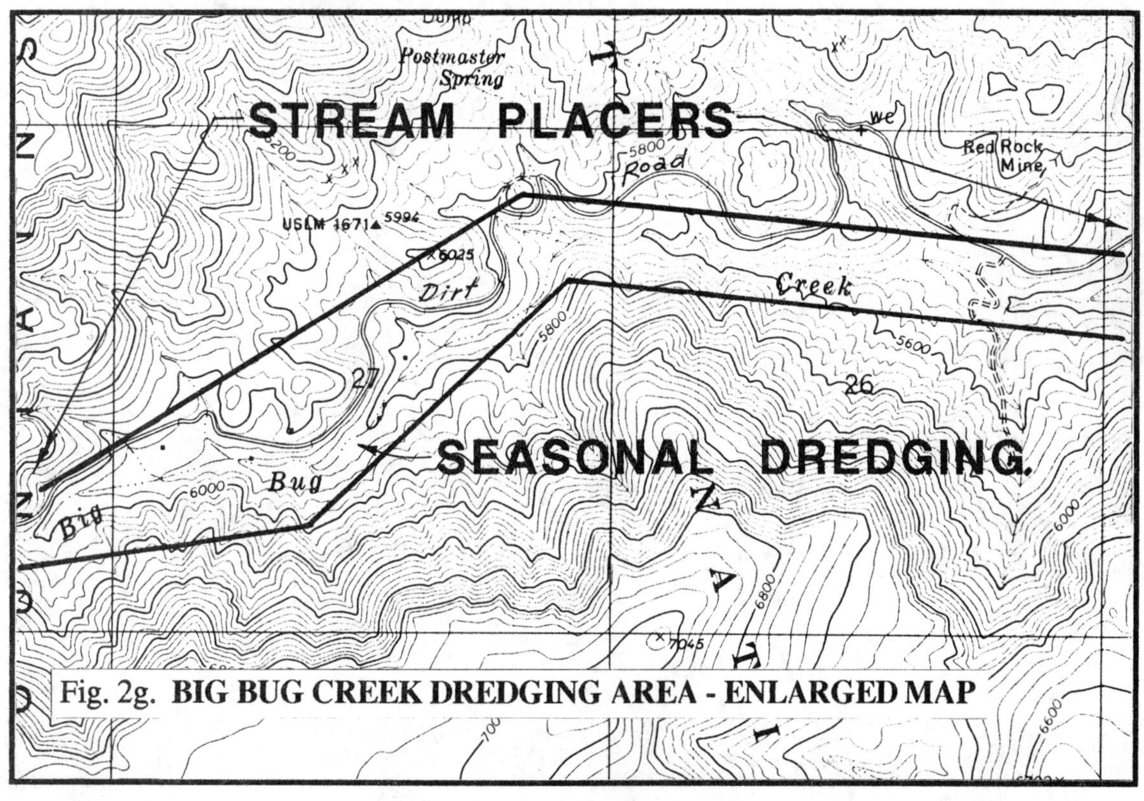

Fig. 2g. BIG BUG CREEK DREDGING AREA - ENLARGED MAP

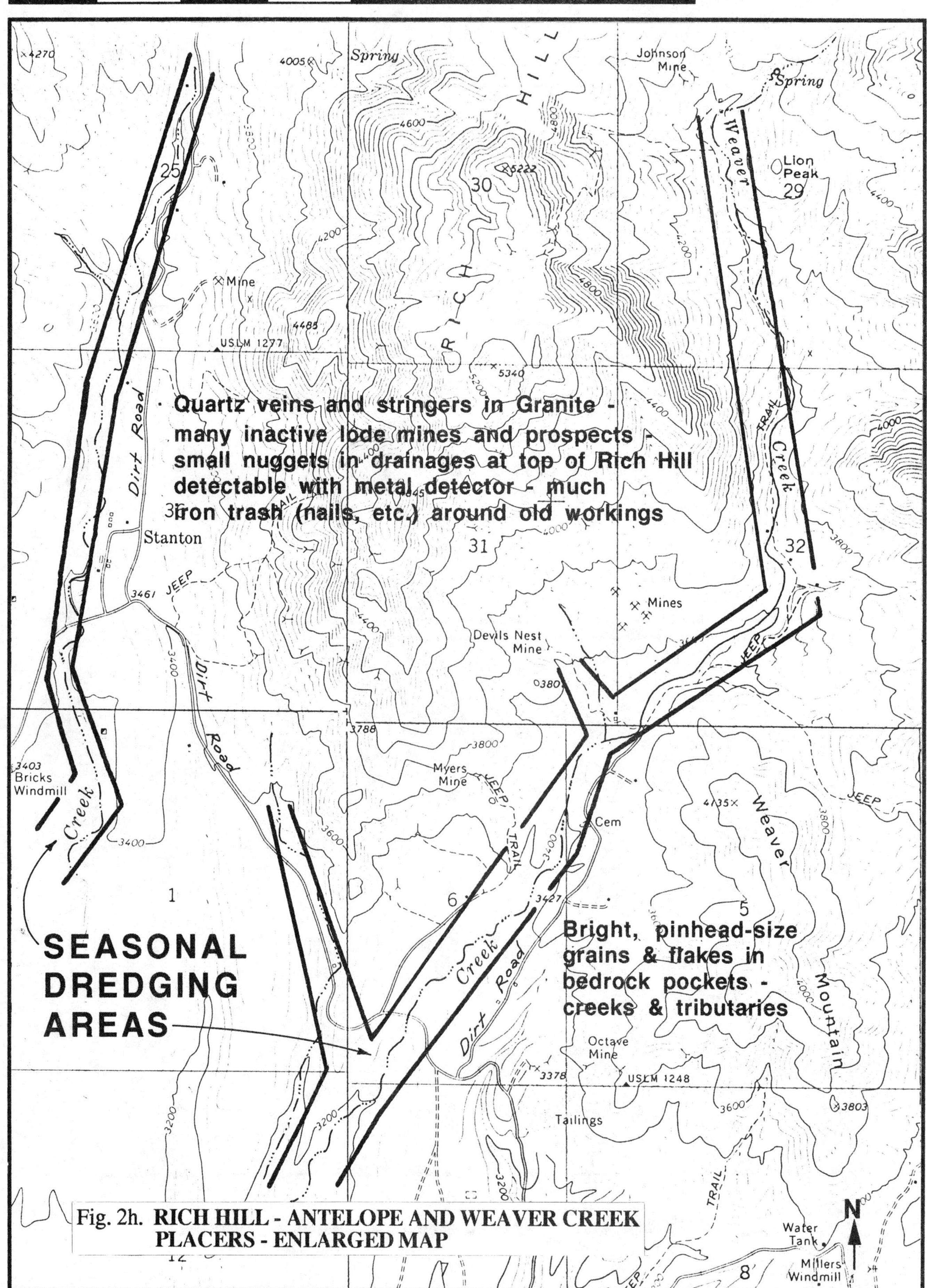

Fig. 2h. RICH HILL - ANTELOPE AND WEAVER CREEK
PLACERS - ENLARGED MAP

SCALE 1: 24000 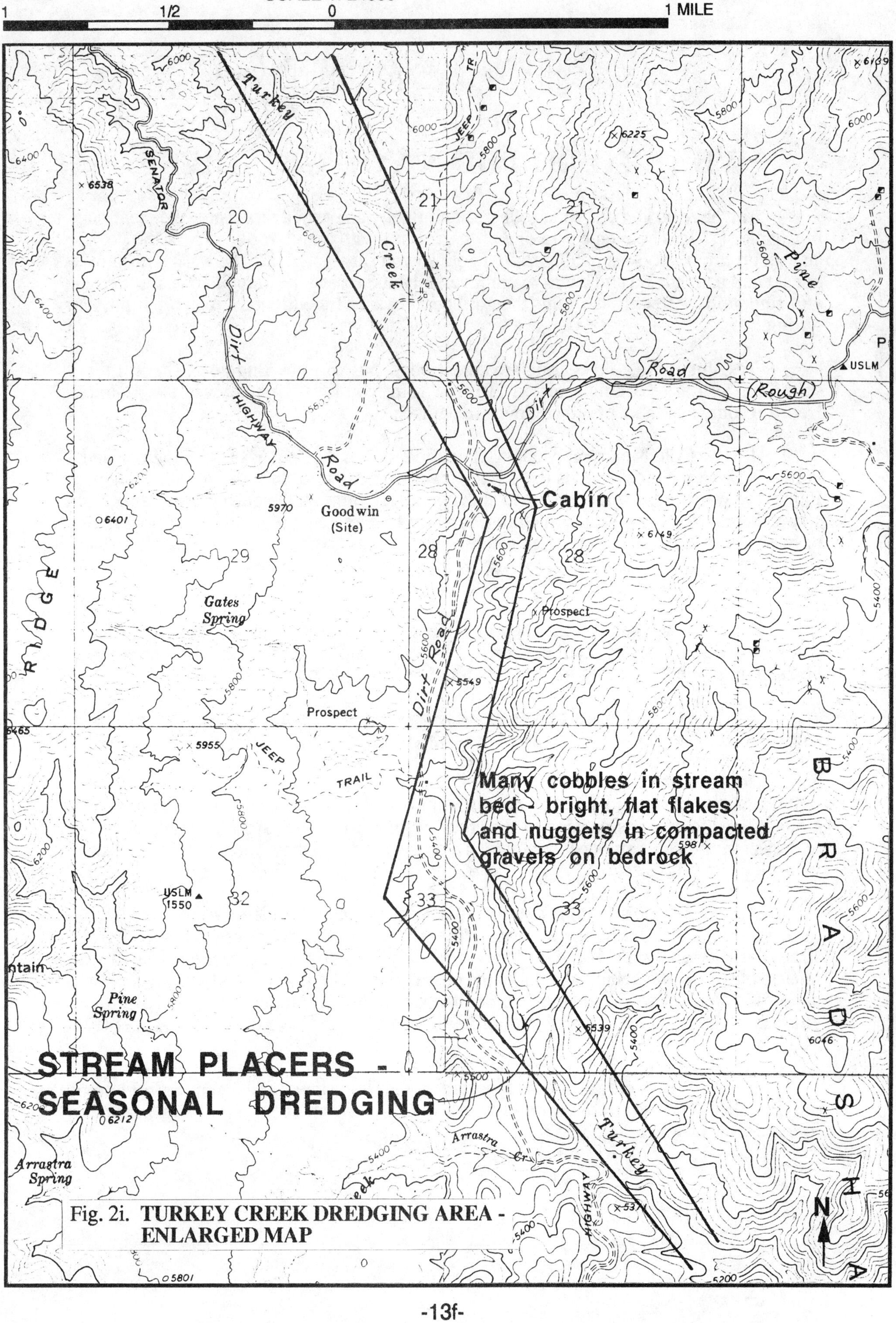
1
1/2
0
1 MILE
6000
Turkey
SENATOR
Creek
6400
× 6538
20
6000
21
JEEP TR
6000
5800
× 6225
21
5800
Pine
5600
6000
× 6109
5800
P
USLM
Dirt
HIGHWAY
Road
Road
Dirt
(Rough)
5600
5600
5600
Cabin
5600
5970
Goodwin
(Site)
28
28
× 6149
6401
29
Gates
Spring
× Prospect
Dirt Road
RIDGE
Prospect
× 5549
6465
Many cobbles in stream
bed – bright, flat flakes
and nuggets in compacted
gravels on bedrock
5955
JEEP
5800
5800
TRAIL
5400
598 ×
R
6200
A
USLM
1550
32
33
33
5400
5600
D
ntain
Pine
Spring
× 5539
6046
I
STREAM PLACERS –
SEASONAL DREDGING
5600
S
6212
Turkey
Arrastra
Cr.
Arrastra
Spring
5400
eek
HIGHWAY
× 5371
N
H
Fig. 2i. TURKEY CREEK DREDGING AREA –
ENLARGED MAP
5200
A
5801

CHAPTER III

WHAT TO DO WHEN YOU GET THERE

You have picked out your gold area and finally arrive at a promising site mentioned in Chapter II. It is convenient to one of the washes with ample room to park the vehicle and set up camp. You are itching to find gold so you rush down into the wash with shovel, bucket and pan in hand and begin to dig feverishly and screen the gravel. You are sure you will find gold. After all, isn't this the gold bearing area I mentioned? You are right, and wrong.

Yes, this is the gold bearing area mentioned, but will you find gold? I don't think so, unless you are extremely lucky. To find gold, you must first analyze the situation. Here are some facts about gold you must remember.

1.   Gold is extremely heavy. Pure gold is about 19 times heavier than an equal volume of water. Since water is the prime mover of gold, what does the gold do? It sinks! In moving water, gold will continue to sink downward through the gravel, inexorably, until stopped by something. That, something, can be a true bedrock, a false bedrock of caliche, or hardpan, or clay. In desert areas where water flow is sporadic, the gold distributes unevenly throughout the gravel. However, gold usually occurs on bedrock. Therefore, you should direct your efforts toward getting to bedrock.

2.   When gold reaches bedrock it takes much force to move it. If it sinks into a crack or crevice, it may stay there for a very, very long time measured in thousands of years, or until the savvy gold hunter pries it out of its hiding place.

3.   Gold is lazy. It takes the easy way down the watercourse. It hugs the inside bends of the creeks and washes where the water slows down (Fig. 3A). Imagine a gold nugget sitting on bedrock and the force of the flow of water reaches a point where the gold dislodges. The water does not push it, but rather, pulls it along. The gold moves as if attached to a string and a person downstream is reeling it in like a fish. It will move in straight lines to each inside bend of the creek. It won't meander sinuously with the creek or wash. If the creek or wash straightens out, the gold will stay fairly close to the sides where the current is slacker.

4.   Any obstructions to the flow of water such as boulders, tree roots, logs, upturned bedrock and the like will create eddy currents on the downstream side of the obstructions (Figs. 3B, 3D, 4B, 4C). In these relatively quiet conditions, the gold will tend to drop out and settle behind and under the obstructions. So, if you see a large boulder in a part of the creek or wash, and this boulder happens to be on bedrock, you may have found yourself a bonanza. Get behind and under this boulder and take out all the gravel (Fig. 3B).

5.   Where creeks and washes widen, the flow slows, and again the gold drops out (Fig. 3E). In these relatively quiet spots the gold will remain for a long time, or until a forceful flow of water dislodges it.

6. Where tributaries connect with the main watercourse a slowing of water will take place just downstream of the tributary, on the tributary side. Here, again, the gold will drop out (Fig. 3A).

7. A terrace of gravel lines the watercourse on one or both sides (Fig. 3F). It may be anywhere from a few feet to several feet above the present creek or wash. There may be several such terraces, and each characterized by a level top. What can this mean? It can mean that in earlier times the stream flow was at the level of the terrace. The flow may have continued for thousands of years, then the watercourse shifted and cut downward into its present location. These terraces have been a rich source of gold, particularly if laid down over bedrock. You will be wise to test out the gravel.

8. Tree roots (Fig. 3D). These are marvelous traps for gold. Not only do the larger roots create eddy currents on the downstream side, but the fine roots act like a net and catch and hold the gold. My son and I had tremendous success on the Gravel Gertie Claim when we suction-dredged into the roots of a tree growing on one side of the creek (Fig. 12). Its roots reached into bedrock, which was a joy to behold. Even the roots of bushes and grasses can trap gold; so don't ignore these.

Generally, look for obstructions to the flow. If you happen to dig onto broken bedrock, with many cracks, you are into good terrain (Fig. 3C). Ignore exposed and scoured bedrock. This indicates a swift flow. Likewise for bedrock that forms a series of waterfalls even though the watercourse may be dry. The gold does not usually stay here, and if it does, grinds down. You should be looking for any slowing action of the stream flow, or obstructions, or both.

You have walked the creek or wash, upstream and down and have found some promising spots. Now, you should set up for a test. If this is a dry wash, get out your shovel; buckets; screens; trowels; pick; screwdriver, etc. If you have a drywasher, bring it down to the spot where you will be working. Dirt is very heavy so you don't want to be carrying it a long way to your drywasher. If you want to test the gravel in your gold pan, it will be easier to take your gravel to where you have the water and not vice versa.

You need to determine how thick is the paydirt. This is the gravel most likely to carry gold and you need to know just how deep it is to bedrock. You must shovel off the top loose layer of sand and gravel. Remove this barren material wide enough around to permit you to work in the excavation. Toss it onto the side of the bank opposite to where you will be working. Get rid of any large rocks you come across. If you feel that you might toss out a nugget along with the waste, then check out each shovel full with your metal detector, if you brought one along. However, this is a very remote occurrence and I generally don't worry about it.

As you dig into the gravel, a change in their make-up should occur. They become firmer, more compact, coarse, and their color may change from the light, sandy one to a darker, brown one. They may also contain some very dark, sometimes black, heavy rocks. These are magnetite and are indicators that you are entering paydirt. From now on, you must remove and screen each shovel full of gravel into your bucket. Check each screenings' waste carefully for nuggets too large to pass through the screen.

As you dig deeper, the gravel may become very firm and compact. This indicates that the gravel has been relatively untouched by the movement of stream flow for a very long time. Look for dark brownish to reddish brown dirt apparently cementing the gravel. This indicates much iron oxide (rust) and is a good sign. Usually, the gold is in this dirt.

Continue digging downward until you reach bedrock. Hopefully, the hole will not be too deep. If you have heeded my advice in Chapter II on avoiding much overburden, you have also avoided digging a hole that reaches to China! I try to limit my dry excavations to no more than two feet deep to bedrock, because of the amount of time and energy expended to dig out that amount of gravel. However, I know that very large nuggets may be deeper.

Once you reach bedrock, you have exposed not only the depth of the overburden at that point, but also the thickness of the paydirt. Now, you should enlarge the hole. Take off the barren overburden around it until you reach the compacted gravel I talked about, and again dig down to the bedrock. As you enlarge the hole, you begin to see what the bedrock does. It will undulate, dip, and form pockets (Fig. 4F). You may see cracks in it, broken and shattered parts. Now, you are into some serious mining.

Collect and save every bit of gravel on the bedrock. When you have enlarged the hole enough to stand in it, get out your whisk broom and sweep it clean; save the sweepings and put them into your bucket. You must clean out *every* crack down to its very bottom, no ifs, ands, or buts. If need be, pry the crack open with a pick, sledge and chisel, or pry bar. You want the bedrock so clean that you could eat off it if necessary. By so doing, you will not let any gold escape you.

At some point you will run out of paydirt. You certainly don't want to dig needlessly. If the obstruction is a large boulder, then you don't want to go upstream of it more than a foot. If you are working the inside bend of the creek or wash, then you don't want dig beyond where the curve of the bend straightens out, both upstream or down, unless the creek or wash widens out below the bend. You will have to pay close attention to the character of the gravel. As you run out of paydirt the gravel begins to change in color and texture becoming more like the surface waste that you tossed away. However, if the bedrock indicates that it is plunging, you may want to follow it some more to see if the character of the gravel changes back to the paydirt color and texture. Sometimes, the bedrock will dip and form a pocket (Fig. 4F). The localized concentration of compacted gravel and dirt in it can hold good gold values. Once you reach bedrock it is difficult to leave it, so you may want to continue mining it. Again, you will have to pay close attention to the gravel and test them every so often to see if you are still getting gold.

If you have a drywasher testing is relatively easy (see Chapter IV). If you have only a gold pan it will be a little more work. Select one shovel full of gravel taken off the bedrock and screen this into your gold pan. Pan down this material to the black sand concentrate. When done correctly, any gold collected with the gravel is immediately evident in the pan. If you do not see any color you are probably out of the paydirt.

If you are going to tackle one of the terraces or high benches presumed to be a remnant of earlier stream flow, you should look for obstructions. Large boulders (as big as a car and bigger) are very good, as they have probably remained in position for a very long time. Trees are good, too, but we measure their lives in decades, not millennia. The approach to mining the terraces remains the same; look for the compacted gravel and try to get to bedrock.

If the gold hunter likes to hike, there are washes hardly touched by any mining. Soap Creek, where I found my first nugget, is such a wash. I worked a small tributary characterized by a bedrock of schistose rock. This rock has a habit of forming many cracks and fissures. In this case, the cracks ran perpendicular to the creek flow and the overburden was no more than a few inches. What I did was to sweep off the overburden and that exposed the cracks. Then, systematically, I went on to clean out a number of cracks using a spoon, screwdriver and whisk broom. This is a method of mining known as sniping. The paydirt went into my knapsack, and when it got full it was time to go.

A short hike took me to a place on the main creek where rain water had become trapped in depressions on a large area of exposed bedrock. It was here that I panned out the paydirt and found my nugget, which was about the size of my little fingernail. If you work in dry creek beds such as these, don't expect to see the nuggets in the dirt. Dust and dirt cover them and they appear like the rest of the material. You have to pan the paydirt to reveal the gold, and you cannot be sure that you even have any gold until you pan out the dirt.

If you decide to do this kind of prospecting, you should have with you some basic tools. These are a folding shovel; trowel; whisk broom; spoon; screen; screwdriver to use as a probe; knapsack to carry the paydirt, and your gold pan. Additionally, you will have to carry your drinking water and food. All this can get rather heavy, and unless your physical condition is good, not recommended. A knapsack full of paydirt is very heavy so you don't want to be carrying it around for very long. Find some water holes, puddles or even cattle tanks in which to pan out your dirt *before* you return to your camp. For the hardy individual, this kind of prospecting can prove very profitable.

***

Soap Creek, Arizona, 1963

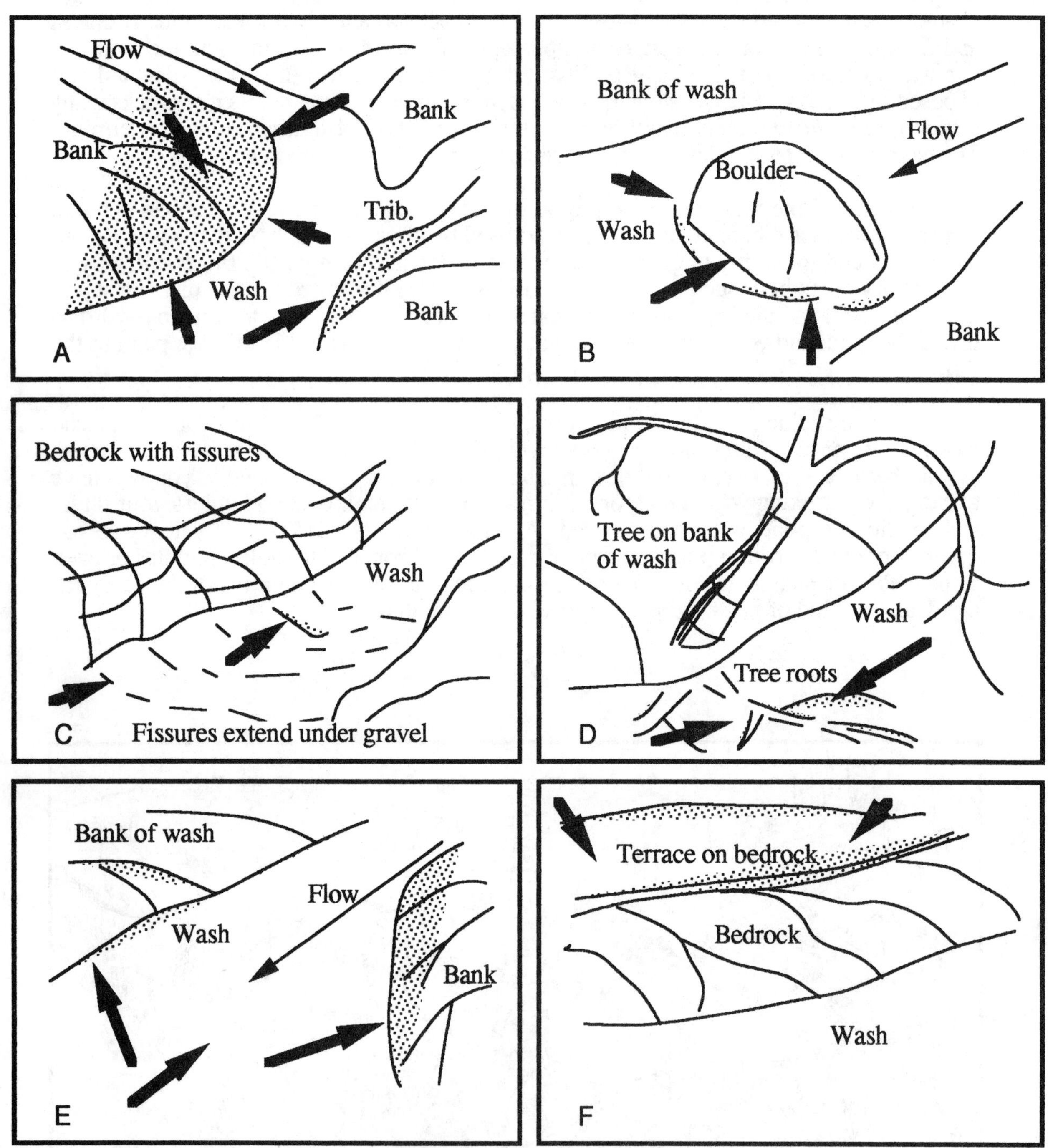

Fig. 3. WHAT TO LOOK FOR WHEN YOU GET THERE
(Heavy arrows point to likely concentrations of gold). A, Inside bend of wash; B, Boulder on bedrock; C, Fissured bedrock extends under wash; D, Tree roots trap gold; E, Widening of wash or creek; F, Terrace gravel on bedrock may indicate ancient level of streambed.

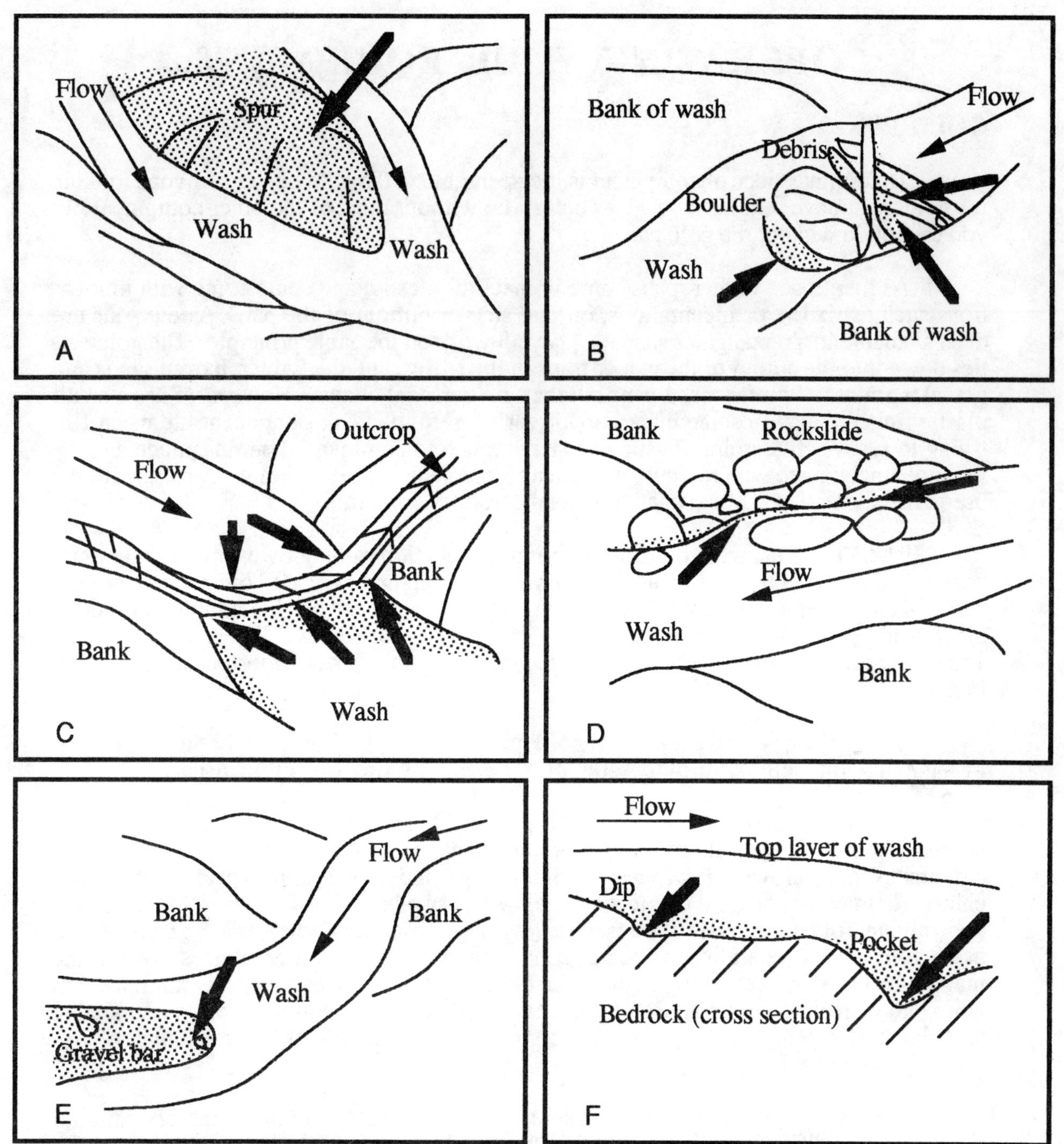

Fig. 4. WHAT TO LOOK FOR WHEN YOU GET THERE
(Heavy arrows point to likely concentrations of gold). A, Spur at intersection of two washes; B, Debris pile against boulder; C, Outcrop cutting across wash; D, Rockslide or boulder pile on one side of wash; E, Gravel bar where steep dip flattens; F, Cross section through bedrock showing dip and pocket.

CHAPTER IV

OPERATING YOUR EQUIPMENT

GOLD PAN:

This simple piece of equipment is indispensable to the gold hunter. Anyone looking for gold must have it in his arsenal. You can do without most of the other equipment but you cannot do without the gold pan.

As mentioned earlier, pans come in metal or plastic; and some come with innovations such as riffles, or indentations, on one side or bottom of the pan. A few pans are even available in a rectangular shape! They all work on the same principle. The gold settles down into the corner of the pan or traps in the riffles, and the lighter, barren sands and gravel wash out. I prefer metal pans but that is a personal choice. However, I use a small plastic finishing pan in some instances for very careful panning of concentrate when I'm trying to recover fine gold. Plastic pans have much to commend them inasmuch as they will not amalgamate with mercury and are non-magnetic. They will not interfere with the use of metal detectors. Metal pans and metal detectors don't mix.

The idea of using a gold pan to go into production mining owes much to the misconceptions most of us have about gold mining. We picture a grizzled old prospector with pan in hand washing out beautiful gold nuggets. This imagery originated with the Forty-niners and perpetuates today in books, magazines, TV and movies. This is not to say that you can't mine gold using a gold pan, but the method is inefficient, laborious, slow and tiring.

The primary purpose of a gold pan is to test your gravel to see if you are getting any color, and to separate the gold from the concentrate.

The serious gold hunter will use his pan in the field to test likely spots for gold. If he gets good color, he will go on to more efficient methods of gold recovery using the equipment that follows: drywasher; suction dredge and sluice; gold wheel or other machine. He may use the gold pan to separate the larger pieces of gold after accumulating a good amount of concentrate. If he faces a very large amount of concentrate he may opt to use a gold wheel, separating table, jig, concentrating bowl, or other devices now on the market for this purpose. Mining gold is hard work. What you want to do is minimize the labor and increase the output. So, if you can afford it, buy the equipment that is available to make this happen. Otherwise, go the poor man's route and use only a pan, but don't expect to get much gold this way.

You use the gold pan with or without water. I do not recommend the dry panning method as it is difficult to perform well and more than likely will result in a loss of gold. Use the wet method only. If you have no water available for wet panning, then save your paydirt until such time and place that you can wet pan it.

Many books on gold mining describe various techniques of gold panning. It is as hard to describe as it is to perform well, rather like rubbing your belly and patting your head at the same time. After a while, with practice, it comes naturally.

First, you should have a large container of water in which to pan. A 20 gallon wash basin, filled not more than a few inches, is fine. Before you begin panning, put a drop of liquid dish washing detergent into the water. This will help remove the oils from your fingers, gold pan and the gold itself. Otherwise, the fine gold will float and wash out of your pan.

Second, clean your pan thoroughly with hot soapy water and rinse well. If your pan is metal, heat it over the burner of your kitchen range to burn off the oils that may have penetrated into the metal. After you have done this you are ready to begin panning. A small stool to sit on will be helpful.

Fill your pan about one-half full of paydirt. You may want to pre-screen your material through a 1/4 inch mesh, but this is not necessary. If you do screen it, check the material not passing through the screen for any nuggets. Wash this material with water first to reveal any dirt covered nuggets.

With both hands, grip each side of the pan and lower the entire pan, contents and all, into the water. Holding the pan under water, thoroughly knead the dirt to break up any clay balls and to wet all material. Remove any large rocks, but check first to see that is what they are; you wouldn't want to toss away a good-sized nugget. Then, slowly lift the pan while at the same time moving it in a circular motion. This swirls the material and allows the heavier gold to settle to the bottom of the pan. Then, in one smooth motion, begin to raise the pan almost out of the water but letting the edge farthest from you to dip down.

Tilt the pan; and with the farthest edge just below the surface of the water, continue the circular motion. This brings up the lighter material that washes over the edge while the gold remains in the groove at the bottom of the pan. Do this until you have mostly black sand and very little light-colored sand and rock remaining in the pan. Now, lift the pan slowly to an almost horizontal position with just a little tilt remaining and gently swirl the water in the pan. As you continue the swirling motion, you will see that the black sands wash backwards. If you have any gold, it will reveal itself immediately as the black sands wash away.

If you are fortunate in seeing some sizable gold, you may want to remove it from the pan. Here is where you use your tweezers to pick it up. Keep your oily fingers out of the pan and off the gold. Pick up the gold only over the pan and drop it into your plastic container or vial that you are also holding over the pan. Don't carry your gold so that if dropped it falls to the ground, or else you may never find it again. Wash any remaining small flakes and specks in the pan into a small bucket, along with the black sands, for later amalgamation with mercury.

After you have panned several pans of paydirt, or concentrate, you will begin to understand the real purpose of the gold pan.

DRYWASHER:

The principle used by the drywasher for separating gold is rather simple. Natural gold, in air, is about 5 times heavier than the gravel in which it occurs. Because of this, the use of small puffs of air passing upwards through the gravel allows the gold to settle downwards. You may trap this gold in your drywasher behind a series of cross bars called riffles while the lighter, barren material blows away. The gravel put through a drywasher must be dry, the drier the better, bone-dry if possible. Depending on the nature of the gold-bearing gravel, it is sometimes possible to put through material that is barely moist, but not recommended. Moist gravel may clog the pores of cloth covering the riffle tray rendering it ineffective. It may make the gold adhere to particles of dirt and be lost with the tailings. Drywashers are excellent devices for recovering gold in dry desert areas but rain, and moist gravel, is their undoing.

There are several kinds of drywashers available. Some are bellows operated and some are continuous forced air. The bellows operated ones may be manual or motorized. The manual machines may use a pull cord to operate the bellows or a hand cranked pulley wheel system. The motorized machines use a small electric motor run off a utility battery. Apart from convenience, the operation of the electrically motorized machine is no more efficient than a manual one. The continuous blower machine uses a gasoline motor and impeller to drive air through the cloth of the riffle tray. This is a large and expensive piece of equipment requiring at least two persons to operate effectively. In remote desert areas, where services are non-existent, the use of machines reliant on an external power source, other than human muscle, is risky. If something goes wrong, you are out of luck and your gold-hunting trip is a bust. I subscribe to the KISS principle: Keep It Simple Stupid.

The manually operated drywasher consists of a double A-frame acting as legs. These support the hopper, riffle tray, and bellow box with bellows. A crankshaft, with piston, moves the bellows up and down. A small pulley wheel connects to the crankshaft, and a larger one, with a crank handle attached, connects to the smaller one with a rubber V-belt. As the large wheel turns, it turns the smaller one that, in turn, moves the crankshaft up and down. The piston, connected to the crankshaft and the bellows also moves up and down, thus creating the puffing of the bellows. If you use a wheel of 12 inch diameter for the larger one and a wheel of 2-1/2 inches for the smaller, for every revolution of the larger wheel the smaller wheel will turn 4.8 times. This translates into 4.8 up and down movements of the bellows for every revolution of the larger wheel. If you maintain one revolution per second of the larger wheel, the puffs through the riffle tray will be 4.8 per second. Any faster than this and you may begin to lose your fine gold.

The riffle tray captures the gold while allowing the lighter, barren material to slough off. It is a rectangular frame, usually made of wood, with crosspieces (riffles) of wood, 3/4 inch square, spaced regularly about 4 inches apart. A muslin cloth stretches across the full area of the frame underneath the riffles. This cloth has many small holes in it that allow the air from the bellows to puff through. A screen of hardware cloth underneath the muslin keeps the muslin from sagging under the weight of the gravel. Underneath each riffle is a flat strip of wood, or metal, so placed that it extends beyond the riffle on the uphill side about 1/4 to 1/2 inch. This strip creates a dead air space behind the riffle which traps the gold. Once into this dead air space, the gold is difficult to dislodge. The riffle tray attaches to the bellow box by clasps, hook eyes, or other means, and is removable. The joints between the riffle tray and the bellow box must be air tight, and using some sort of weather-strip accomplishes this. The riffle tray and bellow box set at an incline so that paydirt fed at the upper end can travel stair step fashion to the lower one and out as tailings.

In the process, as the bellows puffs the paydirt, any gold present in the gravel will work its way downwards to the muslin, then downwards into the dead air space against the riffle where it will remain. However, as the riffles load up with black sand concentrate, incoming gold may ride over the riffles and lost with the tailings. To prevent this, empty the riffle tray regularly.

I design and custom-build my own drywashers (Figs. 5 & 6), and tailor these to my needs and to the specific nature of Arizona's gold-bearing dry washes. They are lightweight and you can carry them to the spot where you will work. The hopper holds a full 4 gallon bucket of paydirt, and has a mechanical feed for metered flow of material. Other features include a vibrating action of the riffle tray, and a thumping action to settle the gold on the muslin. Recovery is excellent.

Suppose you have filled several buckets of paydirt from the hole you dug and want to process this through your drywasher. If you have a manually operated drywasher, fill the hopper to its capacity. For the machine to work properly, you must cover the riffle tray with dirt. Take some from the wash and fill the tray level with tops of the riffles. Turn the crank on the larger wheel to check the puffing action. Now, open up the hopper to let out some paydirt, and continue cranking the larger wheel. As the bellows move up and down, you should see dust, puffing upwards from the tray, and the gravel sort of hop up and over each riffle. Meter out the gravel from the hopper so that you don't overload the riffle tray, and maintain a steady action of the bellows. Don't pump too fast or you may lose the fine gold. Remember, about one revolution per second for the larger wheel is just about correct.

When the hopper is empty, stop pumping. Close the hopper door and refill the hopper with more paydirt. Continue to repeat this process until you have fed about four buckets of paydirt through the hopper. Some washes have much black sand, so you don't want to overload the riffles. If you keep on pumping after the hopper runs dry, and if it's done gently, much barren material will slough off. However, don't overdo it or you may lose your fine gold with the tailings.

Detach the riffle tray, and being careful not to spill its contents, dump the material out into a large rectangular plastic washbasin or other container. This is your gold-bearing concentrate. This is the stuff you save for later processing. If you are careful in the manner in which you dump the material into your container, you just might see some gold on the black sands. However, most of the time, dust covers the gold and you will not see it.

Replace the riffle tray and repeat the whole process again. At the end of the day, the basin will contain quite a bit of concentrate. **Treat this as money.** You may now pan the concentrate and separate the gold, or save it for separation later at your convenience. If you accumulate much concentrate, it may pay for you to invest in a gold wheel, concentrating bowl, or other recovery device.

SUCTION DREDGE:

A suction dredge consists of a sluice box with riffles, header box, liner, clear plastic suction hose, pressure hose, foot valve with hose, pump and gasoline engine. It operates in such a way that water from the pump, under very high pressure, feeds into the suction nozzle, thereby creating a vacuum. As the suction nozzle moves over the gravel, the vacuum created sucks up all material capable of passing through the suction hose, and delivers it to the header of the sluice box. The sluice box inclines in such a way that the water and gravel delivered to the header box will wash over the riffles and out the end of the sluice. Any gold present in the gravel gets trapped in the header and behind the riffles. Cleanup of the header and riffles is usually after one hour's running time of the engine, but may occur sooner, or later, depending upon the gold content of the gravel. The material taken from the header and riffles is concentrate, or cons, for short.

The availability of water restricts the use of suction dredges. They are very popular on the West Coast, especially in California's mother lode area. Another type of dredge is the power jet. This dredge works best in rivers and streams with a good flow and ample depth. In Arizona, with its gold-bearing streams of low flow, and shallow, intermittent creeks, the suction dredge is the most practicable type of dredge. With this dredge, it is possible to use the suction nozzle effectively in as little as an inch of water!

Suppose you want to operate a suction dredge on one of the gold-bearing streams in Arizona. Again, the approach is similar to that outlined in Chapter III. You should be looking for obstructions, and a slowing down of the water.

In suction-dredging, the sluice box sets up so that the waste, or tailings, goes downstream. Therefore, set up the sluice box so that you aren't going to recycle the tailings back into the sluice. Start your dredging well downstream from the obstruction, or slackening of the stream, and work the suction nozzle upstream. However, before you do this, set up the pump motor and foot valve. Since the motor for a suction dredge is gasoline powered, you should have brought with you gasoline in metal containers, flexible pouring spout, engine oil, and an old tire off a compact car. Then, do the following.

1.   Check the oil level in the engine's crankcase and make sure that it's filled to capacity. An engine sitting unused for a long time should have the oil drained. Refill with fresh oil before starting out on your gold trip.

2.   Check the gasoline tank for contamination from dirt or water. It's a good idea to run the tank dry at the end of every gold trip to avoid bad gas on your next trip. If it's clean, add fresh gasoline up to the fill line. Be very careful with gasoline. No cigarettes or open flames near it, please. Try not to spill any gasoline onto the ground or into the stream.

3.   Select a spot, preferably level, on the bank close to the stream or even in the shallow part of the stream, and set down the old tire here. On it, place the engine. The tire will act as a vibration isolator and will also elevate the engine above the stream.

4.   Attach the foot valve hose and foot valve to the pump's intake, and screw on the pressure hose to the pump leaving it just a little loose (do not screw it on tight, yet). Screw the pressure hose coupling onto the suction nozzle and be sure that it is tight.

5.   Set up the sluice box downstream of where you plan to dredge. Incline the box at approximately one inch per foot to allow the water and tailings to flow out without losing the gold. Gather up some large rocks upon which to set the sluice box at this proper pitch. Place a rock on the header to prevent the sluice box from being dislodged.

6.   The suction nozzle should have the suction hose attached to it. I never disconnected mine. Now, attach the suction hose to the header box of the sluice.

7.   Prime the pump before you start the engine. Ideally, the water will be of sufficient depth so that you can place the foot valve in a small wash basin or bucket submerged in the stream. This will prevent sand from entering the pump and eventually ruin-

ing it. Unfortunately, the streams in Arizona are not ideal, and you may have to place the foot valve directly on the stream bed. If the stream or creek is too shallow, you will have to dig a hole large enough in which to put the foot valve.

8.   Prime the pump by raising and lowering the foot valve rapidly, or by moving it upstream and down. As you prime the pump, water will start to flow out the loose coupling of the pressure hose. Start the engine. If the pump has a good prime, a strong flow of water will come out the loose coupling. Tighten down the coupling. The pressure hose will become firm and water will start to flow through the sluice. If the pressure hose does not become firm and no water flows in the sluice, **shut the engine down immediately**. The pump has lost its prime, and running it dry can damage the seals. Re-prime the pump, and make sure that you have loosened the coupling from the pump just a little bit. When water flows through the sluice, set the engine speed to about one-half. You are ready to begin work.

Now, clear away the overburden. Using the suction nozzle accomplishes this best. You may use a sharp-pointed shovel, but the compacted sand, gravel and large rocks may give you fits. Take down the top layer of barren gravel in as wide an area as possible to give you room to work. Don't dig a post hole in the stream. Suction dredge the paystreak in even layers. After each layer, shut off the engine, and examine the header box and riffles for any gold. You must pan the material in the sluice to determine if you are reaching gold. Continue working upstream, and downward toward the bedrock, at the same time. The character of the gold-bearing gravel will change if you are into a good paystreak. They are very firm and compact and may take a sharp-pointed bar to break up. You may come across many cobbles and boulders; remove these from the hole. Toss these to one side of the stream, or other, but not in the direction of where you plan to work.

Operating a suction dredge takes much practice. If you plunge the suction nozzle into the gravel it is liable to become plugged. Then, you will have to unplug it with a rod, or strike it sharply with a rubber mallet to dislodge the plug. You can largely avoid this by using a gloved hand held at the mouth of the nozzle, and immediately remove any large rocks being sucked into the nozzle. If you use your bare hand to do this, as I have on occasions, the incoming rocks can really hurt the fingers. The use of a glove may help to alleviate this. In spite of this precaution, plug-ups do occur in the suction nozzle, or the hose, from time to time. Just hit the nozzle, or the hose, at the point of plug-up with a rubber mallet. This should dislodge the rock, which most likely is causing the problem.

Between shutdowns of the engine, check the gasoline and oil levels. You can run the engine until the tank goes dry, then refill it for more running time, but **you should not let the crankcase oil go dry**, unless you want to buy another engine.

If you run the motor too fast, then, the swift flow of water in the sluice may wash out the gold. If the sluice box has too much of an incline, then, the gold may not stick around and may wash out with the tailings. Do not allow large rocks to accumulate on the riffles, thereby altering the even flow of water; remove them immediately. If the flow in the stream cannot support the capacity of the pump, it will lose its prime as the stream flow draws down. In some cases, you may have to recycle the water if the stream flow is insufficient. Watch the foot valve closely and keep it cleared of leaves, roots, sand, etc. In spite of these annoyances, suction-dredging is one of the best ways to get gold. A small dredge can move an enormous amount of material in a few hours, and on a gold-bearing stream, the ability to move material is the key to getting much gold. Unfortunately, Arizona's streams don't have sufficient flow to support the larger dredges (4" and bigger) which can move a tremendous quantity of material. The intermittent flow in some of the smaller creeks may be totally inadequate for all but the smallest suction dredges (1-1/2 inch back-packer type).

I mentioned recycling the stream water where flow is insufficient to support the pump's demand. Here is one way to accomplish this.

1.   Gather rocks, large and small (usually available in abundance in the area), and construct a small dam across the stream, laying in the larger rocks, then smaller ones, and filling the spaces between them with sand and mud. This will give you an impound-ment. To enlarge its depth, shovel out some of the stream bed.

2.   Place your sluice box *upstream* from where you are working and to one side of the stream. This is contrary to what I said earlier about it being downstream so you don't recycle your tailings, but because of the insufficient stream flow, you make this ex-ception.

3.   Construct another smaller dam around the tail end of the sluice box. This allows the silt and muddy water to settle before it reaches the larger impoundment.

You should now have enough water for a full day's operations, but it will become very muddy and may silt up the foot valve.

If you begin suction-dredging early in the morning, you may find the flow in the stream much stronger than in the afternoon. This may puzzle you. What accounts for it is that the trees and plants growing along the stream will transpire more water from their leaves during the heat of the day, and their roots will take in more of the stream water. What had been a sufficient flow earlier to support your pump becomes totally inadequate later. In that case, you must construct an impoundment for recycling water, or halt opera-tions.

Suction-dredging performed by two, or more persons, is best. One person wields the suction nozzle and the other tends the sluice box and pump motor. As the person oper-ating the nozzle works the stream, material passing through the sluice must not build up. The second person will keep the larger rocks moving through the sluice. In a long opera-tion, you will have to remove the tailings at the end of the sluice, or they will build up and wash into the dredging hole. The second person shovels, or rakes, this material down-stream. In this way, he keeps the waste out of the dredging hole.

Suction-dredging is cold, wet work and hard on the lower back. You should have wet gear to wear while you are in the stream, and a back-saver to prevent the aches and pains that may result from handling a heavy suction nozzle for several hours. In the months of May and June, the water may warm up enough for you to go into it in wearing only shorts and a pair of sneakers; I did. Suction-dredging is not for everyone. However, for the serious gold hunter, it is a good way to get gold.

GOLD WHEEL:

I described the operation of this device in an earlier chapter (Chapter I). It is a won-derful time saver for separating gold from concentrate. You can also use it to process classified (screened), raw gravel. I used my gold wheel regularly in this fashion while mining the Gravel Gertie Claim in the dry times and was very successful in recovering large nuggets. It uses a 12 volt utility battery to operate the motor. With a 20 gallon wash basin holding 5 gallons of water for recycling, you can take this device to the gold fields for direct processing of paydirt. What you get in the final process is not concentrate, but gold, gold separated from the dirt.

The material fed into this device must not exceed 1/4 inch in size, and the smaller the better. This rules it out for the larger nuggets. In some areas, nuggets exceeding 1/4 inch are rare, so this device might work out well. It has four drawbacks. 1. It requires water to do its job; 2. It is slow; 3. The spray bar plugs up easily (and must be unplugged with a bent wire such as a paper clip); and, 4. It is cumbersome and takes up much room. Its use is better at home where the availability of water and electricity is not a problem. In spite of the drawbacks, this device recovers all but the smallest gold. To recover almost all the fine gold, you have to use a concentrating bowl. The recovery of the finest gold is one of the challenges facing miners today, as no device, no matter how efficient, will recover all the gold.

METAL DETECTOR:

You can find gold nuggets using this highly sophisticated, electronic metal-locating device. There are many models available. Some machines are strictly for finding gold nuggets. Others are for several purposes but can locate gold if they are of high quality, and the manufacturer expressly warrants that they have circuitry designed to detect gold nuggets. Stay away from the poor quality metal detectors usually sold in discount department stores, as they won't do the job.

If you are going to invest in a metal detector, get only the best machine that is capable of tuning out hot rocks, and with automatic ground balancing. Otherwise, you may get frustrated using a machine without these features in the highly mineralized areas where you find gold. A high quality metal detector is not inexpensive. Prices, at metal-detecting and prospecting supply stores start around $500 for a strictly gold-hunting machine, and go up from there. A good all-around machine will allow you to use it for other things besides gold, such as, coinshooting, treasure hunting, etc., whereas, the strictly gold-hunting one is effective for doing one thing only, finding gold. The choice is yours. Your local dealer is very knowledgeable in the use of metal detectors and will gladly help you to make the best choice suited to your needs and budget. He will instruct you in the proper use of the machine you select and provide excellent warranty and repair service if required.

Searching for gold nuggets with a metal detector takes much skill, practice and *patience*. Some people are quite proficient at it. However, you may search for hours and never find a nugget. You may search for days, or weeks, and never find a nugget. If you do, it may be small. If you are really lucky, it may be big. The important thing is not to get discouraged and impatient. Nuggets *are* elusive. Learn how to use your machine. Then go out to the gold fields and use it! The more you use it the better you become with it.

I use my metal detectors to check out the tailings from my drywasher, the screenings too big to pass through the sieves, and the shoveled out overburden, for any nuggets that may have escaped recovery. As you can see, all this material is in one spot and easily checked out; looking for nuggets with a metal detector any other way requires much hiking, and time. This is fine if you like to hike, which is excellent exercise, and at some times of the year when the gravel is too moist from rains, this allows you to continue to look for gold.

If you elect to metal detect for gold, go to the drywashing gold fields. You may have some luck searching the smaller washes and rivulets feeding the washes. Always keep in mind that gold likes to hang up behind obstructions. You may take your cue from what I said about how I use my detectors and confine your searches to the tailing piles, screenings, and overburden piles left by other drywashers. Many drywasher operators are careless about how they operate, and many do not bring metal detectors with them to check out the waste material. You could do very well raking out these piles and running your metal detector over them. If you can find an area where the old timers drywashed (difficult, as most of the piles have become obscured with time) you could do very, very well. The machines that the old timers used were very inefficient and much gold went off with the screenings and tailings. Pay particular attention to the coarser screenings. Check these out for nuggets that were too large to pass through the screens the old timers used. The shoveled out overburden piles may contain nuggets; check these out, too. The old timers did not have metal detectors then and relied strictly on eyesight. Many a dust-covered nugget escaped detection. You may understand how easy it is to miss seeing gold when you find your own dust-covered nugget with a metal detector.

Suppose that you have your detector with you and you want to check out one of those drywasher piles. Turn it on. Set the ground balance adjustment (a machine with automatic, self-adjusting ground balance will do it for you), away from any metallic objects or hot rocks, and adjust the threshold sound (some machines are silent operating). Use no discrimination, or at most, "1" on the discrimination level. Set to maximum sensitivity, or just below the level where the machine's sound starts to waver. You should have with you a small plastic gold pan, a digging tool (trowel, small pick, or small shovel), an ore bag (or knapsack), and a plastic film container for your nuggets. Now, you are ready to check out the pile.

1.   Slowly move your detector's search coil over the pile to detect any surface nuggets.

2.   If you do not get a response, scalp off the top layers of the pile and spread this material around. You have to get into the center of the pile but not all at once, just a layer, or so, at a time. Again, move the search coil over the pile.

3.   Continue these operations until you either get a response or none at all, and you have flattened and gone over the entire pile.

4.   If you get a response, pinpoint the location with the search coil. Use your trowel or shovel to open up the material at the point of response and check again. If you moved the target, there will be no response at the hole. Now, you know that it is in the material to the side of the hole. If you have not moved it, there will be a response over the hole. Take your trowel or shovel and dig under the location of the response. Place this material into your plastic gold pan.

5.   Move the search coil over the material in the pan. If you have done this properly, the target should be in the pan and you will get a response. Try to pinpoint its location. Reach in with one hand, free of rings, watch or other metallic object, and grab the dirt. Laying your machine on its side, pass the material you now hold in your hand across

the search coil. If you have a response, the target is in your hand.

6.   Now, divide the material by passing one-half of it to the other hand, and pass each hand over the search coil. Discard the material that doesn't elicit a response from your detector. Continue doing this until you just have a small amount of material remaining in the hand eliciting the signal. Visually inspect this material for the nugget. If you see a nugget, that's wonderful! Remove it and place it in your film container, being very careful about holding the container over your gold pan should you drop the nugget. If you don't see it, put this material into your ore bag, knapsack, or other safe container for later panning. Not all responses will be gold. They could be pieces of lead from bullets, nails, wire, etc. However, the response from these is usually loud and sharp. Gold gives a softer response, unless you hit a big nugget. You can take out ferrous (iron) objects by attaching a magnet to the handle of your digging tool and passing it over the material in your pan. What remains may be brass (bullet cartridges), lead (bullets), aluminum, or gold.

OTHER OPTIONS:

I have covered the basic equipment necessary to the all-around, serious gold hunter. As you become more proficient in gold hunting, you may want to increase your reward. You are about to cross the line from the weekend gold hunter, the recreational miner, to the professional. You are about to enter the realm of big bucks and environmental problems.

1.   You should have a claim to work. This gives you the mineral rights to the exclusion of all others. You can own the claim outright by locating it yourself, or by purchase, or by leasing it from a valid claim holder.

2.   Your claim must have gold in paying quantities. In other words, after you have deducted all expenses and sell the gold, will you have a profit left over?

3.   The U. S. Government requires you to pay a number of fees and post restoration bonds to restore the area after mining. Various kinds of state and local agencies may also require you to pay fees, and secure mining permits before you begin operations.

4.   You may engage in long legal battles with environmentalists strictly opposed to mining in any way, form or shape. The portrayal of the large-scale miner today is that of a rapist of the environment; you will find little public support for your cause.

In spite of this, you may decide anyway that you want to go ahead. There are several machines available to help you.

1.   Front end loader.

2.   Backhoe.

3.   Bulldozer.

4.   Trommel with grizzly and self-contained recyclable water reservoir.

5.   Placer jig.

6.   Gasoline-powered drywasher with a battery of four or more riffle trays that

operate simultaneously.

7.   Feed conveyer.

And, more. Advertisements for large scale mining equipment appear throughout the issues of the *California Mining Journal* and you may want to get professional advice from a mining engineer, geologist, attorney, etc., before embarking on this venture.

If this is not the route you want to go, then you might want to remain just small enough in your operation to avoid many hassles. You still must have some property to work, but if your operations are small and don't disturb the environment in a major way, you could do quite well. You need to keep down your expenses on equipment because these are the major cost items. If you have decided to go this route, you have already acquired the expertise to make decisions about equipment purchases.

***

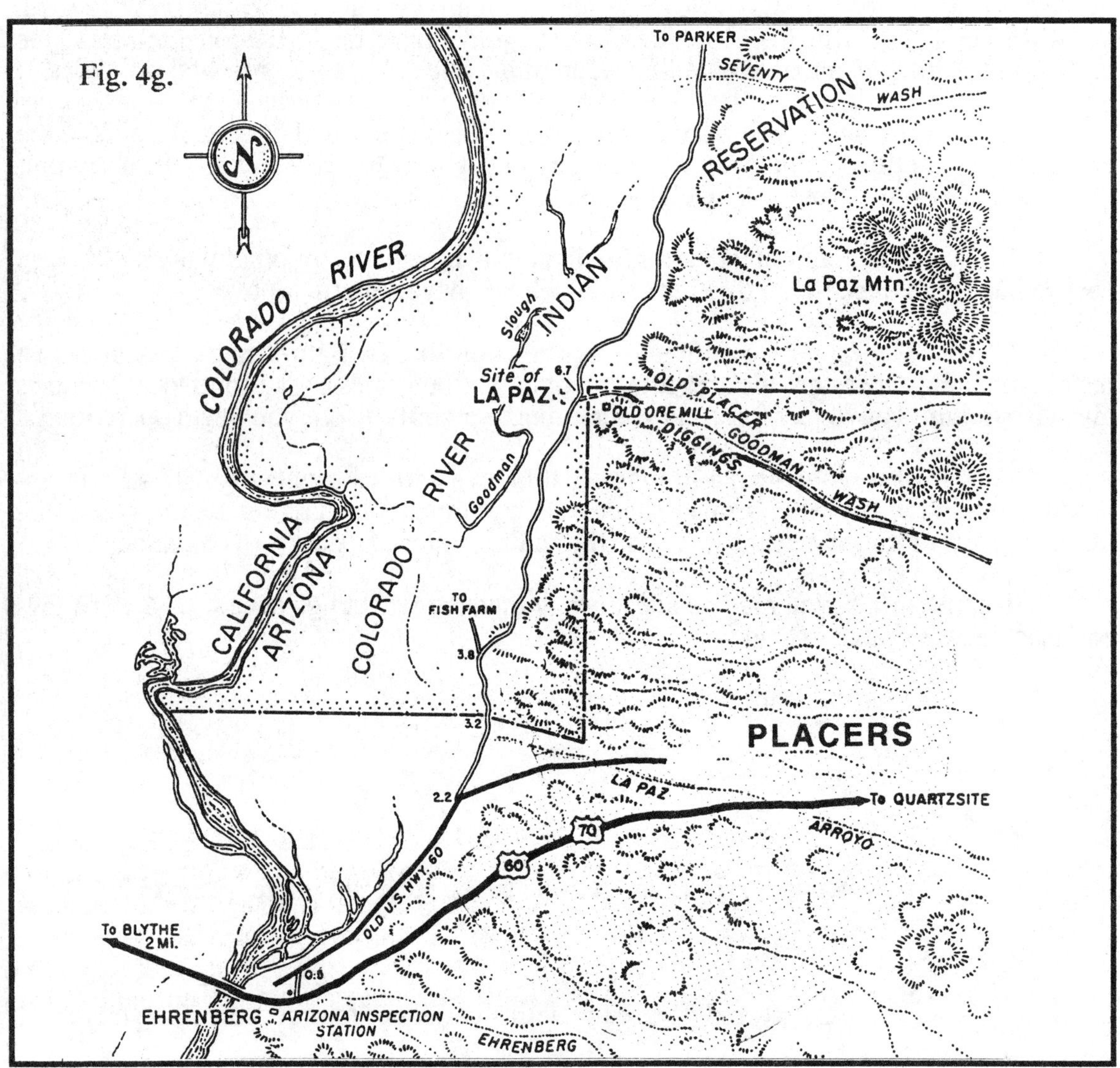

Figs. 5 - 10. **OPTIONAL EQUIPMENT FOR THE SERIOUS PROSPECTOR.**   5, Drywasher, pull cord-operated bellows, two views (custom designed and built by the author); 6, Drywasher, crank-operated bellows, two views (custom designed and built by the author); 7, Portable Screen (custom designed and built by the author). Screened gravel slides down chute by gravity into plastic bucket; 8, Suction Dredge, 3-inch size (photo courtesy of Gold King, Placer Equipment Mfg., Inc.); 9, Gold Wheel; 10, Metal Detectors (upper: all-around machine; lower: strictly for gold).

Fig. 11. Author drywashing in the Las Guijas Mountains, Pima County, Arizona. A, Set-up on the inside bend of a wash and start of excavation; B, Screening gravel with the portable screen; C, Excavating to bedrock (overburden removed and digging into paydirt); D, Sweeping the bedrock (clean out *every* crack, fissure and dip). A gasoline-powered vacuum cleaner, that the author also uses, is an excellent device for cleaning bedrock. Paydirt thickness (6", arrow) is exposed. Bedrock is humped and plunging to the right; E, Start of drywashing (note full hopper); F, Cranking the drywasher (note tailings pile and buckets of paydirt); G, Dumping the riffle tray concentrate into a plastic wash basin for later panning (riffle tray is emptied after no more than four full hoppers have been drywashed). Cotton gloves protect hands.

Fig. 12. Roger Wielgus is operating his 3 inch suction dredge on the Gravel Gertie Claim, Ash Canyon, Huachuca Mountains, Cochise County, Arizona, May 1983.

He is working the suction nozzle in the roots of the tree just visible above and to his right. The overburden at this spot is approximately two feet and the large boulders throughout are typical of Ash Creek. Sluice box with gold pan on the header box is in the right foreground; pump motor at extreme left. Large rocks sucked up by the nozzle must keep moving through the sluice and not allowed to settle on the riffles. Otherwise, the smooth flow of water will disrupt causing a loss of gold. Suction-dredging by two persons is best. One wields the suction nozzle and the other keeps the sluice clear. In a water recycling operation the sluice tender also regulates the amount of water returning to the excavation, thus permitting the person on the suction nozzle to work in very shallow water. Low water flow is characteristic of intermittent streams in Arizona and many will be dry by May.

The large boulder just visible above the pump motor rested on a 4 inch wide, deep bedrock crack that extended completely across the creek from bank to bank. Subsequently, he dredged this crack and completely removed the compacted pay dirt by using a sharp pointed bar. It proved to be a bonanza with many large nuggets recovered. Large boulders weighing many tons such as these rarely move by stream flow on intermittent creeks as the force of water is insufficient even in times of flood. As a result the gold that accumulates behind and under these boulders tends to remain there indefinitely and builds up substantially over the years. In the above situation we got a bonus as a part of the crack extended into the tree roots! Pay close attention to the header box, however, whenever you are working in tree roots. The smaller, fibrous roots sucked up by the nozzle will get tangled up in the header screen and clog it.

Fig. 12a. The author and son suction-dredging on their Gravel Gertie Claim, Ash Canyon, Huachuca Mountains, Cochise County, Arizona, April 1985. A, Recycling low stream flow back to the excavation (note smooth flow through sluice); B, Roger tossing rocks to waste pile on right (rubber gloves protect hands); C, Working in the excavation among the tree roots (excellent gold situation); D, Roger panning the header concentrates (thumb of right hand scrapes off waste material frompan); E, Going for the gold is fun! Author, and son panning (low stream flow is no problem for a 3-inch suction dredge if you recycle the water); F, Using the pan correctly with just the right amount of tilt (note sluice, riffles and classifier screen).

-33a-

Fig. 13. The author and son suction-dredging on their Gravel Gertie Claim, Ash Canyon, Huachuca Mountains, Cochise County, Arizona, April 1985. A, Roger on the suction nozzle; B, Ronald tending the sluice; C, Cleaning out the header. Most of the larger gold is found here, the smaller gold in the first two riffles below the header. But, save the concentrates from all the riffles, which will contain the smallest gold; D, Panning the header concentrates; E, Swirling the black sands and checking for color; F, Gold!

The paydirt here occurs mostly on the bedrock as a dark brown, gritty material containing many small pebbles of magnetite and hematite, but very little black sand. It also forms the "cement" between the many cobbles and boulders lying just above bedrock. Pushing the suction nozzle into it gave a distinctive "feel" and it was not necessary to see the business end of the nozzle. Instead, we watched the material passing through the clear suction hose for the "look" that is associated with this feel. This way of mining results from many years' experience on a suction dredge.

Drywashing in the Las Guijas Mountains, Pima Co., Arizona. Author's set-up on a bend of a small wash. Rear: Portable Screen; Front: Drywasher.

CHAPTER V

SEPARATING THE GOLD

Before you begin separating the gold from your concentrates you should first learn to recognize what it looks like. Natural gold nuggets, flakes and colors (fine gold) all have similar characteristics. These are color, weight, malleability, ductility and insolubility. Let's discuss each of these characteristics in turn.

1.   Color. Gold in a pure state is brassy-yellow. Natural gold is virtually never pure but alloyed to some extent with other metals, primarily silver and copper. Silver alloyed with gold will give it a lighter color and a high percentage of silver to gold will give the gold a silvery color with a yellowish tint. Copper alloyed with gold will give the gold a reddish tint. The color of gold, when seen under different light conditions, be it shade or sunlight, does not vary; it will always appear the same. It will also appear metallic. Mica, often mistaken for gold will change its color and appearance under different light conditions. The same is true for iron pyrite (fool's gold). When you see bright yellow specks in your black sand after carefully panning down your concentrate, and their color doesn't vary, you are probably looking at gold.

2.   Weight. I talked about this earlier. A given volume of pure gold is about 19.3 times heavier than an equal volume of water. This is very important. In the panning process, you are taking advantage of this characteristic of gold to separate it from the lighter, worthless material in your pan. If you have panned your material carefully, you will have washed away most of the light, worthless minerals such as mica and pyrite. What you should have left in the bottom of your pan will be black sand (magnetite) and gold. Black sand is obvious. Then, what are those bright yellow specks and flakes that resist budging when you swirl the water in your pan? Probably, gold.

3.   Malleability.   You can hammer gold flat without it cracking or breaking. You can also hammer pure gold so thin that light will shine through it, but it will not break. A sharp knife will cut gold. If you hammer or cut mica and pyrite, they will break, crack or crush easily.

4.   Ductility. You can stretch pure gold without breaking it, and draw it easily into a wire the thickness of a hair and many thousands of feet long.

5.   Solubility, or should I say insolubility. Gold is unaffected by ordinary acids. It will not dissolve in nitric, hydrochloric or sulfuric acid alone. However, if you mix one part of strong nitric acid to two parts of strong hydrochloric acid you get an acid that will dissolve gold, Aqua Regia.

If you find some bright yellow flakes and nuggets in your panned concentrate, you probably have gold. However, you don't want to verify this by pounding them with a hammer, thus destroying all collector value the natural gold may have. Instead, you can use the nitric acid test. Place a small drop of nitric acid on the suspect gold. If it does not fizz and dissolve, you have gold. The gold will be unaffected by this treatment. Use a plastic pan or glass dish, not your metal pan, otherwise the acid may eat a hole in it. After a while, you will become very proficient at identifying gold by sight alone and not have to resort to an acid test.

I talked about retrieving the larger pieces of gold from the black sand. There probably will be many, many smaller flakes and specks that are virtually impossible to retrieve with tweezers. You shouldn't even try. Instead, rinse out all the material in the pan into a separate small bucket and save it. After you have accumulated a good amount of this material you need to separate the gold from it. The following steps are one way you can accomplish this, and I recommend using a plastic pan.

1.   Empty the contents of the small bucket of gold-bearing cons into a large plastic gold pan. Add water to the pan, almost filling it. To this water add one tablespoon of household lye. Slowly swirl the water and the concentrates for several minutes. The lye will clean the gold of all surface impurities such as oil, dirt, etc. (Note: If you have more cons than the gold pan can handle, a small rock tumbler is an excellent way to perform this step and the following one. Put the contents of the bucket into the barrel of the tumbler, add water to fill halfway, then add the one tablespoon of lye and tumble for about an hour.)

2.   Pour off the lye-laden water (carefully), add fresh water and to it add 1/4 ounce of mercury. Be very careful in the use of mercury. Its fumes are toxic if inhaled and handling mercury with bare skin is very dangerous as it can enter the bloodstream through the pores.

3.   Slowly swirl the contents of your pan that now contain the mercury. As you do this the mercury will grab the gold, every little speck of it. If there is much gold the mercury will get sluggish as it becomes heavy with it. It is amalgamating with the gold, coating it, and binding it together into a ball. If you are using a tumbler, you will have poured off the lye-laden water, added fresh water and the mercury, and tumbled the contents for at least 3 hours. I have a hexagonal-shaped stainless steel bar about 1-1/2 inches thick by 5 inches long that I put into my tumbler barrel to help grind any black sand that may have gold adhering to it, thus freeing the gold.

4.   Carefully pan away the black sand. A magnet placed in a plastic bag and moved slowly over the black sand will pick up the magnetite, making things a little easier. Removing the bag allows the magnetite to drop off without sticking to the magnet. You want to remove as much of the black sand as possible leaving only the mercury. As you slowly swirl the material in your pan, you will see the mercury hang up at the head end. It will look thick; it contains the gold. The freer-running mercury does not have any gold in it. Pour this mercury back into your mercury container, but put the container into another plastic pan before you do this just in case you spill the mercury. In that way, you will contain it and not let it run onto the ground, floor, or whatever, thus contaminating the area. (You may want to buy a mercury press for easier separation of the mercury from the gold. It is available at most prospecting supply shops.)

5.   Flush out the balled up mercury into a small glass container, such as a pudding cup, and add some water. Do not add too much, just enough to cover the mercury ball (amalgam). To this, add a few drops of nitric acid (available in small plastic bottles at most prospecting supply shops). Add the acid until you see a fizzing action in the mercury. The acid is dissolving the mercury and any alloyed silver that are going into solution. The fumes arising from this operation are toxic so do this outdoors, and avoid breathing them. After a while, the fizzing will stop. The fumes will stop generating and you will see in the bottom of your dish a bronze-colored, metallic-looking ball. This is gold! Carefully pour

off the liquid into another pudding cup and neutralize the liquid with baking soda. Gently rinse the cup with water and try not to lose the gold. Allow to dry. Then, using a soft sable hair brush, ease the gold into a small plastic film container or glass vial and save it for later processing.

When you have accumulated a small amount of gold in this manner you may want to convert it into a bright shining button called doré. To do this you need a large block of charcoal, propane torch and powdered borax flux. The charcoal and flux are also available at most prospecting supply shops. The following steps will help you obtain your doré button.

1.	Hollow out a small depression in the charcoal just big enough to contain your gold (Fig. 13a.).

2.	Put your gold into this depression (Fig. 13a.).

3.	Fire up your propane torch (not too big a flame) and play the flame over the gold. Gold has a high melting point so you need to keep doing this until you see the gold become bright red (Fig. 13b.).

4.	Still keeping the flame on the bright red gold, add a pinch of borax and turn up the heat slightly. As the gold heats up it will change color, becoming a bright greenish-orange at the point of melting. All the little particles of gold will fuse, then melt into a bright ball (Figs. 13c. & 13d.).

5.	Remove the flame and let the ball cool down slowly. After it has cooled, you have a doré button. It will *look* like gold, a very bright, shiny, metallic yellow. This gold will be nearly pure except for some alloyed copper and other metals.

If you want to create "nugget" type jewelry, there are a few more steps to perform. You need some sulfuric, or muriatic, acid and your small glass pudding cup. Be very careful with sulfuric acid; it is quite dangerous. It is available at most prospecting shops and from swimming pool chemical dealers. Go on with the following steps.

1.	Half fill the pudding cup with the acid.

2.	While the gold still is in a melted state as a bright ball, quickly remove the flame, pick up the charcoal block and drop the ball of gold into the acid. As the gold cools quickly in the acid it will form an irregular shape, a "nugget." You may want to experiment on how you drop the gold into the acid as this will affect the shape of the "nugget." Each time that you do this step you will obtain a differently shaped one.

3.	Remove the "nugget" with stainless steel tweezers and rinse with tap water. It will be bright, shiny and beautiful and as a pendant, worn from a chain or in any number of ways. You have just created a lovely piece of gold jewelry.

4.	Pour the acid back into its container and rinse the pudding cup under running tap water.

To separate the fine gold from the cons directly, not using the above process, it is necessary that you have at your disposal either a gold wheel or a concentrating bowl. If you buy either of these devices, it will come with complete instructions on how to separate the gold from the black sand. What you have after you follow the instructions to the letter will be clean, natural gold. After drying, put it into your plastic container or glass vial.

You now have achieved your goal. You have gotten your share of Arizona gold. What you do with it is your business. You can sell it to jewelers at a premium price for use as jewelry gold. You can also sell your doré buttons to gold buyers, but you must first have them assayed; this will cost you money. Additionally, the gold buyer will discount the bullion value of your gold 12% or more. Instead, you can do like many gold hunters, **save your gold.**

***

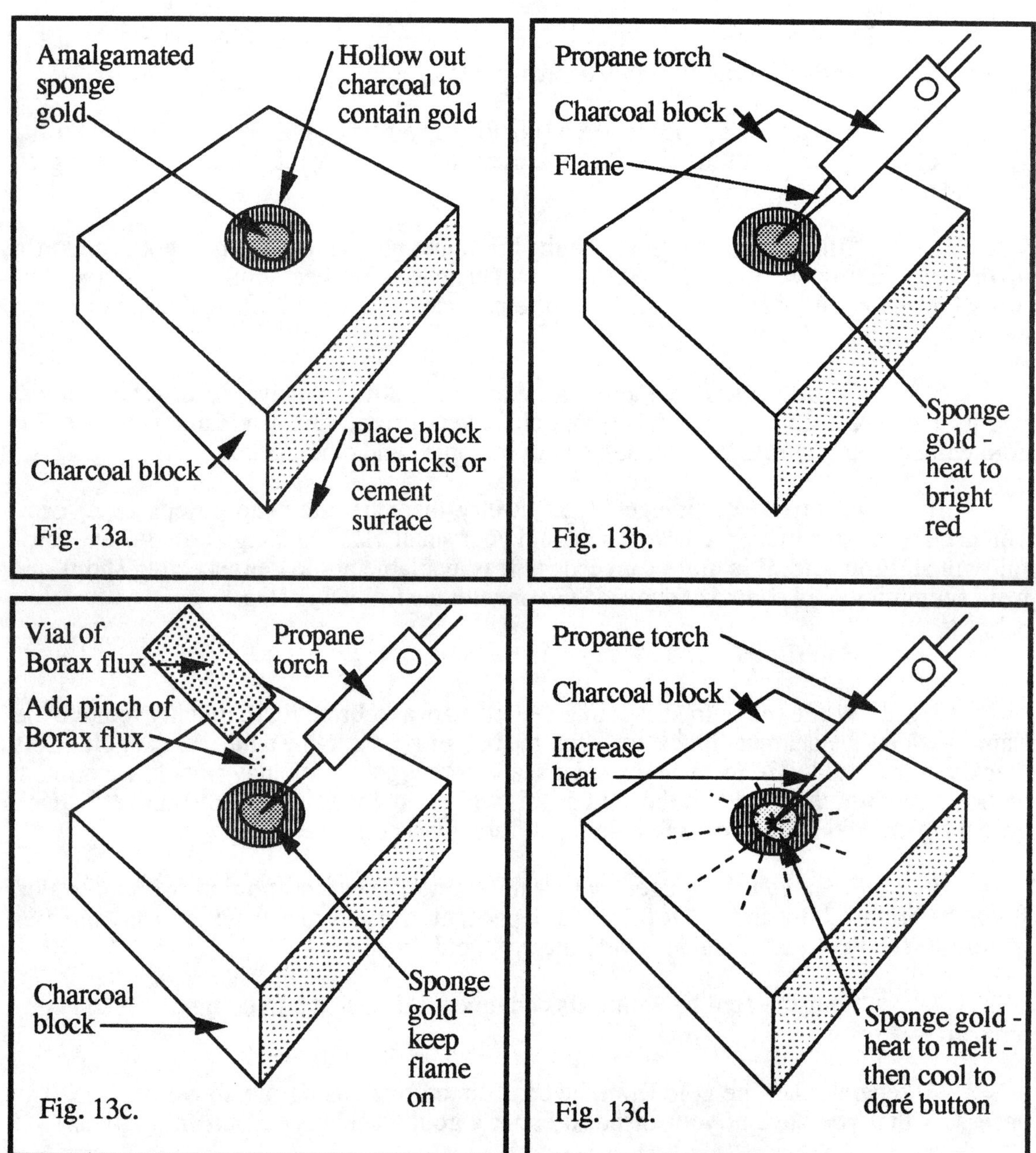

FROM AMALGAMATED SPONGE GOLD TO DORÉ GOLD BUTTON

CHAPTER VI

CLAIMS AND CLAIMING

Throughout this book I mention the word claim, specifically, valid claim. The reader may not be clear what this means and why I advise caution about doing any prospecting or mining in areas where there may be valid claims. I will try to give a general outline regarding claims, the location of claims, who may own claims, etc. I caution the reader that this is not the definitive interpretation of the mining laws and I cannot guarantee the accuracy of the information herein.

For a complete reciting of the federal mining laws, rules and regulations applying to lands in the public domain, the reader should contact:

> Bureau of Land Management
> Arizona State Office
> 3707 N. 7th Street
> P. O. Box 16563
> Phoenix, Arizona 85011

When the gold hunter intends to prospect and mine on Arizona State Lands, to determine the status of the lands, or to secure the latest statutes pertaining to mining on state land, he should contact:

> State Land Department
> 1624 W. Adams
> Phoenix, Arizona 85007

LANDS IN THE PUBLIC DOMAIN:

1. Any citizen of the United States, or anyone who has declared his intention to become a citizen, an association of citizens or a qualified corporation may locate a mining claim upon public domain of the United States. The location of a mining claim by an alien or transfer to an alien is not absolutely void, but is voidable.

There is no limitation on the number of mining locations that the qualified locator may make on federal or state lands.

2. A valid mining claim within the State of Arizona upon public domain of the United States, or on lands where the United States has reserved the minerals, locates either as a lode claim or placer claim. The definition of a lode claim is a "claim upon veins or lodes of quartz or other rock in place." All other "forms of deposit, excepting veins of quartz, or other rock in place . . ." locate as placers.

3. If the land is open to mineral entry the requirements, to locate a mining claim on public domain in Arizona, are

a)   Discovery of valuable mineral in place;

b)   Location by the posting of a location notice;

c)   Recordation of a copy of the location notice and a map, plat or sketch with the county recorder of the county in which the claim is located;

d)   The marking of the boundaries of the claim on the ground; and

e)   The filing of a copy of the recorded location notice and any supplemental information and fee with the Bureau of Land Management in Phoenix, Arizona.

Discovery means the finding and demonstration of the existence of a mineral deposit that you can presently mine, remove and market at a profit. This is the **"Marketability Test."** In essence, to constitute a discovery, you must find minerals in such quantity and quality that, as a person of ordinary prudence, you continue with justification in the further expenditure of your labor and means. You should also have a reasonable prospect of success in developing a valuable mine. It is not enough only to show that minerals exist that would warrant further exploration in the *hope* of finding a valuable deposit. A good rule of thumb: If a claimant cannot derive an income from mining which is at least equal to what he could earn for the same time invested in the labor market, it is unlikely he made a discovery and satisfied the **"Prudent Man Rule."**

The Department of the Interior, of which the Bureau of Land Management is a part, is becoming more stringent in the application of the "Prudent Man Rule." It will apply the rule to determine the adequacy of a mineral discovery in both patent proceedings and in contests challenging the validity of unpatented mining claims.

Location means the discovery of valuable placer material and the erection of a monument or post upon which the locator posts and signs a location notice. The location notice must have the following information on it:

1.   The name of the claim.

2.   The name and address of the locator (the locator can locate for himself, for himself and others, or for others).

3.   The date of location.

4.   The length and width of the claim in feet and the distance in feet from the "location" monument to each end of the claim. This location should conform as nearly as practicable to the United States system of public land surveys and the rectangular subdivision of sections. Strict conformity may not be possible, nor required, where a placer deposit occurs in the bed of a meandering stream.

A placer claim may not include more than 20 acres for each claimant, but there is no limit to the number of claims that an individual or association of individuals may locate. A group of individuals may locate an "association" placer claim with a maximum of 160 acres. Thus, two individuals may locate an association placer claim of 40 acres, three individuals 60 acres, etc., up to the maximum of 160 acres by eight individuals in a single claim. If there are "dummy" locators, the validity of the claim may be subject to challenge.

Record a copy of the location notice within 90 days after the date of location. You should record it with the recorder of the county in which the claim situates. File a copy of the recorded location with the Bureau of Land Management also within 90 days from the date of location. It is wise to record quickly to establish your rights and to serve notice to other possible locators.

When I speak of a claim owner I am usually talking about the locator of a valid placer claim on pubic domain of the United States who acquires the rights to all placer minerals found on his claim. As you can see, only the claim owner has the right to extract valuable placer minerals from his claim to the exclusion of all others. This is very important when you go out prospecting for gold as the area in which you plan to mine already may be under claim. The claim owner has every right to prevent you from extracting his gold if you are on his claim. I hope that he will not resort to violence to do this. If he tells you to leave, do it. His claim may or may not be valid but who wants to get into a fight over it; there are legal means that you can take to resolve the issue.

After you have properly located a placer claim you must perform a minimum amount of labor in developing your claim each year thereafter. You must file with the Bureau of Land Management an Affidavit of Performance of Annual Labor recorded with the county recorder of the county in which the claim situates. You must do this on or before the 30th day of December of each calendar year for assessment work performed during the preceding assessment year. The assessment year is *not* the same as the calendar year. It begins at 12:00 noon on September 1 of one calendar year and ends at noon on September 1 of the following calendar year. To avoid a problem with this perform your assessment work after January 1, but before the last day of August of the calendar year in which you will file your affidavit with the Bureau of Land Management. Make sure that you have recorded your affidavit with the county recorder's office before the last day of August and make sure that you promptly file a copy of the recorded affidavit with the Bureau of Land Management (BLM). **Do not wait until the last day.** If you file with the BLM *after* the 30th day of December, **your claim will become automatically void by operation of law.** Your former claim is now open to location by anyone unless you relocate it before another locator.

When I talk about invalid claims, I am referring to claims that have become automatically void by operation of law. This is due, primarily, to the claim holder not performing his required labor and filing the affidavit as required by law. Each year the BLM declares many claims abandoned as a result of this failure to obey the law. I have included herein at the end of this chapter a copy of a letter from the BLM offices in Phoenix declaring abandoned some lode claims I had. I voluntarily abandoned these claims when I decided not to file the affidavit as required by law.

To determine whether a claim is valid or invalid, you must first search the records at the county recorder's office in which the claim situates to see if the locator has complied with the law. Then, you must check the records either at the BLM offices in Phoenix or at the district and resource area offices. The records at the BLM offices are not always up to date as the work load involving claims is horrendous. You may ask to see the file on a particular claim; each file has a serial number. At the end of this chapter is a detailed step by step process that you should follow when researching a mining claim using BLM records. If the claimant has not recorded the affidavit with the county recorder's office before 30 December of the calendar year, the chances are good that he has abandoned his claim.

If you wish to file a claim, you may obtain preprinted location forms from office supply stores and prospecting supply shops. I have included a sample of a location notice at the end of this chapter.

The appropriate District Ranger of the national forest in which you intend to conduct mining operations requires you to file a "notice of intent to operate" and submit a "plan of operations." You will be wise to contact the District Ranger to obtain the latest rules and regulations before beginning any mining operations that might cause disturbance to the surface resources. They also will advise you regarding searching and removal of small mineral samples or specimens, or on prospecting and sampling operations. Exemptions may exist regarding surface resources not significantly disturbed.

The media has made patenting land a big issue. The impression given is that it is a U. S. Government land giveaway and that anyone can get land by paying only $2.50 per acre. The media gave little attention to the truth of this issue. It is true that claimants may get patent to the claims that they are working but it isn't that simple. The patenting process is very complex and costly to the claimant. Before applying for patent the claimant will have had to expend considerable sums in developing his claim. More than likely when he applies for patent his claim will have to stand a validity test by the U. S. Department of the Interior. Also, it may contest his claim and rule it invalid by application of the "Prudent Man Rule." Instead of obtaining a patent the claimant will probably lose his claim. Sure, there have been abuses of the law; but, should we throw out the baby with the bath water?

ACQUISITION OF MINERAL RIGHTS ON STATE LAND:

The term "state lands" refers to lands and minerals owned by the State of Arizona. Legislation found in Title 27 of the Arizona Revised Statutes, and State Land Department regulations published by the Arizona Secretary of State, governs the acquisition of minerals on state land. The federal mining laws, and the state laws relating to the acquisition of federally owned minerals, do not apply to state lands, except where specifically incorporated.

The State of Arizona requires you to have a state lease before mining any minerals on state land.

The Arizona statutes provide that the discoverer of "a valuable mineral deposit on any state land may enter upon and locate the deposit as a mineral claim." There are two types of state mineral claims, referred to as Type A and Type B.

A Type A claim applies to lodes and a Type B claim applies to placer type deposits. The State does not require you to perform location, or discovery, *work* on Type B mineral claims, but proof of discovery of valuable mineral is necessary.

An alternative to acquisition of minerals by way of a Type A or Type B mineral claim is a state prospecting permit. The permit grants exclusive mineral prospecting rights on state lands during the time the permit is in effect, up to a maximum of 5 years, and the permittee may acquire a mineral lease on lands when he discovers a valuable mineral deposit.

Any citizen of the United States, partnership or association of citizens, or a corporation organized under the laws of the United States or any state or territory and authorized to transact business in Arizona may apply to the State Land Commissioner for a mineral exploration permit on state land. The application must be in writing and contain a description of the land for which the applicant seeks a mineral exploration permit, together with other information required by the State Land Department. Application forms are available from the department. File the application and fee with the State Land Department.

If the applicant qualifies for a prospecting permit, the State Land Commissioner notifies him by registered or certified mail of the amount to pay for the prospecting permit and of any requirement for a bond.

The initial term of a prospecting permit is one year, subject to four annual renewals for an aggregate term not to exceed five years. The permit expires automatically at the end of each annual term unless before expiration the permittee files with the department an application for renewal for the ensuing annual period. At the time of making application for renewal, he must show compliance with certain requirements. These requirements mandate the expenditure of a fixed amount of dollars per acre for "exploration" that can get costly for the small miner. In addition, the permittee must pay to the state an annual rental fee for each acre granted under this permit.

In the case of mineral leases issued for a term of 20 years, there is a fixed amount of annual rental fee for each claim. It is payable in advance at the time of application for the lease and at the beginning of each subsequent annual period. The state also receives a royalty of five percent (5%) of the net value of the minerals produced from the claim. The gross value after processing minus the actual cost of transportation from the place of production to the place of processing, minus the costs of processing and taxes levied and paid upon the production thereof, equals net value.

A state mineral lessee must perform annual labor as required by the laws of the United States upon each claim (lease) or group of claims in common ownership. The annual labor must start at the expiration of one year from the date of location and the lessee must furnish proof of such performance to the State Land Commissioner within 90 days after the expiration of each annual period for performance.

As you can see, leasing state lands for mineral exploration and extraction can get quite expensive for all but the large mining companies.

HOW TO RESEARCH A MINING CLAIM USING BLM RECORDS:

Unpatented mining claim information maintained by the Arizona State Office of the Bureau of Land Management in Phoenix is available on microfiche in the following forms:

1.   Alphabetically, by claim name.

2.   Alphabetically, by owner's name.

3.   Geographic location (section, township, range, meridian, for example, Section 34, T. 10 N., R. 1 W., GSRM).

4.   Serial Number, by the Arizona Mining Claim Number (AMC#) assigned by BLM.

If you wish to research an area for possible mining location you will need to check the following records:

1.   Status records. These records will help you determine if the area of interest is open to location, withdrawn from location or in private or state ownership. They also are a source for the location of patented claims.

2.   Geographic Microfiche Index. It shows claims recorded in each section of a particular township and range. It also shows the claim name; the owner's name; the location date, and the date of the last recorded assessment affidavit.

3.   Mining claim case files. These files are by the Arizona Mining Claim Serial Number (AMC#) and contain copies of the location notices, maps, annual assessment affidavits, and transfers of interest.

You may review the Status Records and all four sets of microfiche (Claim Name, Owners Name, Geographic and Serial Number Indices) at the BLM District and Resource Area Offices listed below. The mining claim case files are available only at the BLM State Office in Phoenix. To begin your research, you will need a map of the area showing township, range and sections. The U. S. Geological Survey Topographic Maps, the Forest Service Maps and the Bureau of Land Management Surface Mineral Management Maps are good sources of information.

If you are planning to visit the BLM offices in Phoenix please bring your map with you. You may buy Surface Mineral Management Maps from the BLM office; however, they are not yet available for the entire State of Arizona. BLM does *not* sell U.S.G.S. Topographic Maps.

The Public Room Staff will be happy to instruct you on how to use the records that are available.

The BLM can provide paper copies of the microfiche indices for $2.00 per page of claims recorded in a particular section, township, and range. You then can order copies of the claim location notices, maps and assessment work for 13 cents per page. Requests for copies of material from the claim files must identify each claim by serial number (AMC#).

The BLM records are open from 9:00 a.m. to 4:00 p.m., Monday through Friday.

BUREAU OF LAND MANAGEMENT DISTRICT OFFICES:

Arizona Strip District Office
390 North 3050 East
St. George, UT 84770
Phone (801) 673-3545

Kingman Resource Area Office
2475 Beverly Ave.
Kingman, AZ 86401
Phone (602) 757-3161

Havasu Resource Area Office
3189 Sweetwater Ave.
Lake Havasu City, AZ 86403
Phone (602) 855-8017

Phoenix District Office
2015 W. Deer Valley Rd.
Phoenix, AZ 85027
Phone (602) 863-4464

Safford District Office
424 E. 4th St.
Safford, AZ 85546
Phone (602) 428-4040

Yuma District Office
3150 Winsor Ave.
Yuma, AZ 85365
Phone (602) 726-6300

CONCLUSION:

In Chapter II, I discussed the probability of many claims on the public domain being abandoned as a result of changes taking place in the mining law. Claimants who do very little actual work on their claims, or have no intention of ever doing any work, but filing each year a "paper" affidavit of labor, hold many of these claims. In the process they have tied up thousands of acres to the exclusion of legitimate prospectors. I think that the passing of pending legislation reforming the mining law and the additional costs imposed (specifically the restoration bonding requirement for disturbance of five acres or less and a rental fee), will prompt many of these so-called miners to abandon their claims. This could open up vast areas to the weekend prospector and serious gold hunter. Only time will tell.

***

Depository Library Network

The Arizona Department of Library, Archives, and Public Records (State Library) and the Arizona Geological Survey (AZGS) have initiated a cooperative program to make AZGS publications more accessible to Arizonans by establishing a series of 14 depository libraries throughout the State (Fig. 14). New AZGS publications (Bulletin, Circular, Special Paper, and Map series) are provided to the State Library, which distributes them to member depositories.

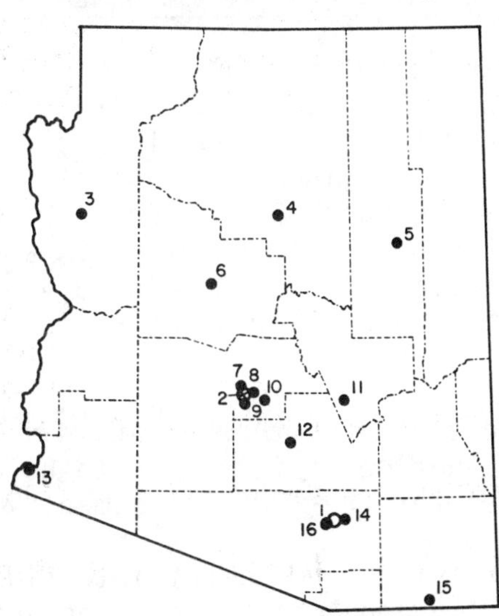

Fig. 14.

1. Arizona Geological Survey
845 Park Ave., Suite 100
Tucson, AZ 85719
2. Arizona Dept. of Library,
Archives and Public Records
1700 W. Washington, State Capitol
Phoenix, AZ 85007
3. Mohave Community College
District Resource Center
1971 Jagerson Ave.
Kingman, AZ 86401

4. Northern Arizona University Library
Government Documents Collection
P. O. Box 6022
Flagstaff, AZ 86011
5. Northland Pioneer College
Learning Resources Center
1200 E. Hermosa Dr.
Holbrook, AZ 86025
6. Yavapai College
Learning Resource Center
1100 E. Sheldon St.
Prescott, AZ 86301

7. Glendale Public Library
7010 N. 58th Ave.
Glendale, AZ 85301
8. Phoenix Public Library
Business and Sciences Division
12 E. McDowell Rd.
Phoenix, AZ 85004
9. Arizona State University Main Library
Government Documents Service
Tempe, AZ 85287
10. Mesa Public Library
64 E. 1st St.
Mesa, AZ 85201
11. Miami Memorial-Gila County Library
1052 Adonis Ave.
Miami, AZ 85539
12. Central Arizona College
Signal Peak Campus
Learning Resource Center
Woodruff at Overfield Rd.
Coolidge, AZ 85226
13. Yuma City-County Library
350 Third Ave.
Yuma, AZ 85634
14. University of Arizona Main Library
Government Documents Section
Tucson, AZ 85721
15. Cochise County Library
Bisbee, AZ 85603
16. Tucson Public Library
200 S. 6th Ave.
Tucson, AZ 85701

United States Department of the Interior

BUREAU OF LAND MANAGEMENT
ARIZONA STATE OFFICE
3707 N. 7TH STREET
P.O. BOX 16563
PHOENIX, ARIZONA 85011
(602) 640-5550

IN REPLY REFER TO:

(922-LC)
0022M
A MC 134528
A MC 137407
A MC 275424

CERTIFIED MAIL--RETURN RECEIPT REQUESTED

September 12, 1990

<u>DECISION</u>

Mining Claimant(s)   Mining Claim(s)
as Shown on the   as Shown on the
Enclosed Sheet   Enclosed Sheet

<u>MINING CLAIMS DECLARED ABANDONED</u>

The Federal Land Policy and Management Act (FLPMA) of 1976, 43 U.S.C. 1744, and the implementing regulations in 43 CFR 3833.2, require an annual filing for all mining claims recorded with the Bureau of Land Management (BLM). FLPMA provides that failure to file evidence of annual assessment work or a notice of intention to hold by December 30 of each year shall be deemed conclusively to constitute an abandonment of the claim and it is void by operation of law. The constitutionality of Section 314 of FLPMA was upheld on April 1, 1985, by the United States Supreme Court in <u>United States v. Locke et al.</u>, 471 U.S. 84, 129 (1985).

The BLM records do not show receipt of either an affidavit of annual assessment work performed or a notice of intention to hold for the claim(s) listed on the enclosed sheet(s) for the 1989 assessment year.

If you did timely file an affidavit or notice of intention to hold with the BLM during 1989, please furnish a copy which shows receipt by the BLM Arizona State Office, (dated and time stamped) during 1989.

Your proof must show the required document was timely filed with the BLM during 1989, otherwise it will not be accepted. The evidence must be received in this office no later than 30 days from receipt of this decision. If the proof is not furnished during this 30-day period, the claim(s) will be removed from our records as abandoned and void.

Alan Rabinoff
Alan Rabinoff
Chief, Branch of
Mining Law Administration

Enclosure

LOCATION NOTICE
(PLACER)

NOTICE IS HEREBY GIVEN that the **HARD TIMES** placer mining claim has been located by Ronald S. & Alyn J. Wielgus, citizens of the United States of America, whose address is 1230 East Placita Del Cervato, Tucson, Arizona 85718. The general course of this claim is east to west and it is situated south of the Old Hat Mining District, Pinal County, Arizona.

This placer mining claim, the name of which is the **HARD TIMES** placer mining claim, situated on the lands belonging to the United States of America, is described as follows, to wit: The North 1/2 of the Southeast 1/4 of the Northeast 1/4 of Section 33, Township 10 South, Range 16 East, G. & S. R. B. & M., Pinal County, Arizona.

DATED AND POSTED on the ground this 26 day of January 1992.

LOCATOR:

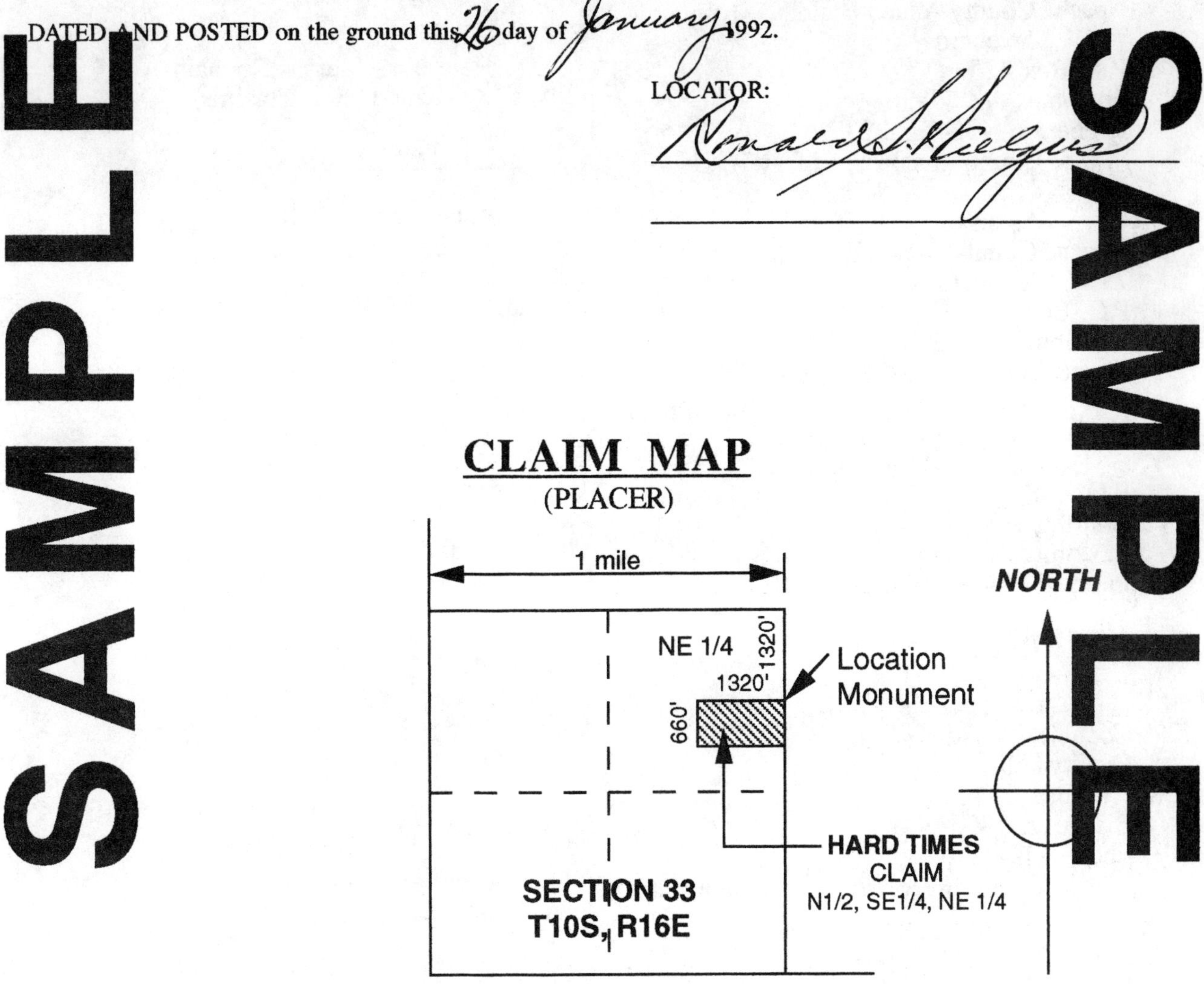

Scale: 2" = 1 mile

1.   The above map depicts the **HARD TIMES** placer mining claim, which is located in Section 33, Township 10 South, Range 16 East, G. & S. R. B. & M., Pinal County, Arizona.

2.   Type of corner and location monuments used are as follows: Stone, 3 feet high.

3.   The bearings and distances between claim corners are as depicted on the above map.

APPENDIX

COUNTY AGENCIES CONCERNED WITH MINING AND MINERAL RESOURCES IN ARIZONA*

APACHE COUNTY

Recorder's Office
Apache County Annex Building
75 W. Cleveland
P.O. Box 425
St. Johns, AZ 85936
Phone 337-4364 Ext. 237
County Recorder - Mary B. Chavez

Assessor's Office
Apache County Annex Building
75 W. Cleveland
P.O. Box 770
St. Johns, AZ 85936
Phone 337-4364 Ext. 224, 226
County Assessor - Anna M. Prentice

Health Department
P.O. Box 697, (Behind County Courthouse)
St. Johns, AZ 85936
Phone 337-4364 Ext. 280

COCHISE COUNTY

Recorder's Office
Cochise Co. Admin. Bldg.
Quality Hill
P.O. Box 184
Bisbee, AZ 85603
Phone 432-9270
County Recorder - Christine Rhodes

Assessor's Office
P.O. Drawer 168, (Behind County Courthouse)
Quality Hill
Bisbee, AZ 85603
Phone 432-9320
County Assessor - Bert E. Merrill

Health Department
619 Melody Lane
Bisbee, AZ 85603
Phone 432-9470
Medical Director - Patrick Quigley

Treasurer's Office
Cochise Co. Admin. Bldg.
Quality Hill
P.O. Box 1778
Bisbee, AZ 85603
Treasurer - Marsha Bonham
(Taxes on patented mining claims)

COCONINO COUNTY

Recorder's Office
Coconino County Courthouse
100 E. Birch St.
Flagstaff, AZ 86001
Phone 779-6585
County Recorder - Helen Hudgens

Assessor's Office
Coconino County Courthouse
100 E. Birch St.
Flagstaff, AZ 86001
Phone 779-6501
County Assessor - Betty Peck

Dept. of Public Health
2500 N. Fort Valley Rd.
Flagstaff, AZ 86001
Phone 779-5164
Director - George Graham

GILA COUNTY

Recorder's Office
Gila County Courthouse
1400 E. Ash
Globe, AZ 85501
Phone 425-3231 Ext. 230
County Recorder - Mary V. DePaoli

Assessor's Office
Gila County Courthouse
1400 E. Ash
P.O. Box 271
Globe, AZ 85502
Phone 425-3231 Ext. 210
County Assessor - Dale Hom

GRAHAM COUNTY

Recorder's Office
Graham County Courthouse
800 W. Main St.
Safford, AZ 85546
Phone 428-3560
County Recorder - Shirley Angle

Assessor's Office
Graham County Courthouse
800 W. Main St.
Safford, AZ 85546
Phone 428-2828
County Assessor - Sam Player

Health Department
826 W. Main St.
Safford, AZ 85546
Phone 428-1962
Administrator - Neil Karnes

GREENLEE COUNTY

Recorder's Office
Greenlee County Courthouse
5th & Webster
P.O. Box 1625
Clifton, AZ 85533
Phone 856-2632
County Recorder - Katie Clonts

Assessor's Office
Greenlee County Courthouse
5th & Webster
P.O. Box 777
Clifton, AZ 85533
Phone 865-5302

Health Department
Greenlee County Courthouse Annex
5th & Leonard
P.O. Box 936

Clifton, AZ 85533
Phone 865-2601
Director - Barbara Brutcher

LA PAZ COUNTY

Recorder's Office
1113 Arizona Ave.
P.O. Box 940
Parker, AZ 85344
Phone 669-6136
County Recorder - Lois Hesse

Assessor's Office
1121 Arizona Ave.
P.O. Box 750
Parker, AZ 85344
Phone 669-6165
County Assessor - Louise Brock

Health Department
702 11th St.
Parker, AZ 85344
Phone 669-1100

MARICOPA COUNTY

Recorder's Office
Maricopa County Complex
111 S. Third Ave.
Phoenix, AZ 85003
Phone 262-3545
County Recorder - Helen Purcell

Assessor's Office
Maricopa County Complex
111 S. Third Ave.
Phoenix, AZ 85003
Phone 262-3406
County Assessor - Ira Friedman

Dept. of Health Services
1825 & 1845 E. Roosevelt
P.O. Box 2111
Phoenix, AZ 85001
Phone 258-6381
Dir., Environmental Services - John Power
Chief, Bureau of Air Pollution Control - Robert Evans

MOHAVE COUNTY

Recorder's Office
Movahe County Courthouse
401 Spring St.
P.O. Box 70
Kingman, AZ 86402
Phone 753-0701
County Recorder - Joan McCall

Assessor's Office
Mohave County Courthouse
401 Spring St.
Kingman, AZ 86401
Phone 753-0703
County Assessor - Kathleen Baker

Div. of Environmental Health
305 W. Beale
Kingman, AZ 86401
Phone 753-0748
Director - Norman Marrah

NAVAJO COUNTY

Recorder's Office
Navajo County Governmental Complex
South Hwy 77
P.O. Box 668
Holbrook, AZ 86025
Phone 524-6161 Ext. 233
County Recorder - Jay Turley

Assessor's Office
Navajo County Governmental Complex
South Hwy 77
P.O. Box 668
Holbrook, AZ 86025
Phone 524-6161 Ext. 203
County Assessor - Randy Andrews

Health Department
Old Courthouse Square
117 E. Buffalo
P.O. Box 639
Holbrook, AZ 86025
Phone 524-6825
Administrator - Maxine McKee

Div. of Environmental Health
Old Courthouse Square
117 E. Buffalo
P.O. Box 639
Holbrook, AZ 86025
Phone 524-6825
Sanitarian - Robert Bixby

PIMA COUNTY

Recorder's Office
Old Pima County Courthouse
115 N. Church
Tucson, AZ 85701
Phone 792-8151
County Recorder - Mike Boyd

Assessor's Office
Old Pima County Courthouse
115 N. Church
Tucson, AZ 85701
Phone 792-8369
County Assessor - Arnold Jeffers

Health Department
Pima County Health & Welfare Bldg.
150 W. Congress, 2nd Floor, Room 237
Tucson, AZ 85701
Phone 792-8261
Director - Guadalupe S. Olivas M.D.

Div. of Air Quality Control
150 W. Congress
Tucson, AZ 85701
Phone 740-8803
Director - David Esposito

PINAL COUNTY

Recorder's Office
Pinal County Courthouse
P.O. Box 848
Florence, AZ 85232
Phone 868-5801 Ext. 391
County Recorder - Kathleen C. Felix

Assessor's Office
Pinal County Courthouse
P.O. Box 709
Florence, AZ 85232
Phone 868-5801 Ext. 361
County Assessor - Paul Larkin

Air Quality Control District
Administrative Building #1
P.O. Box 1076
Florence, AZ 85232
Phone 868-5801 Ext. 486
Director - Martin Godusi
Deputy Control Officer - Suzanne Laughlin

SANTA CRUZ COUNTY

Recorder's Office
Santa Cruz County Courthouse
Court & Morley
P.O. Box 1150
Nogales, AZ 85621
Phone 281-4768
County Recorder - Mary Lou G. Sainz

Assessor's Office
222 Plum
Nogales, AZ 85621
Phone 281-3501
County Assessor - Armando Dela Ossa

Div. of Environmenal Health
1025 Bejarano St.
Nogales, AZ 85621
Phone 281-4901
Deputy Director - Patrick Zurich

YAVAPAI COUNTY

Recorder's Office
Law Enforcement & Administrative Center
255 E. Gurley St.
Prescott, AZ 86301
Phone 771-3254
County Recorder - Patsy Jenney-Colon

Assessor's Office
Law Enforcement & Administrative Center
255 E. Gurley St.
Prescott, AZ 86301
Phone 771-3220
County Assessor - Clair Barnett

Health Department
930 Division St.
P.O. Box 2111
Prescott, AZ 86302
Phone 771-3121
Director - Marcia Gardiallas

YUMA COUNTY

Recorder's Office
168 S. 2nd Ave.
P.O. Box 511
Yuma, AZ 85364
Phone 329-2061
County Recorder - Glenys Schmitt

Assessor's Office
168 S. 2nd Ave.
P.O. Box 1528
Yuma, AZ 85364
Phone 329-2018
County Assessor - Alberta Smith

Public Health Department
201 S. 2nd Ave.
Yuma, AZ 85364
Phone 329-2230
Director - Larry A. Leach

Div. of Environmental Health
201 S. 2nd Ave.
Yuma, AZ 85364
Phone 329-2230
Supervising Sanitarian - Mike Poradek

* Current as of July 1991

***

GLOSSARY

AFFIDAVIT OF ANNUAL LABOR - Document attesting to the performance of annual labor or improvements on the claim, signed by the claimant and notarized. Must be recorded with the county recorder of the county in which the claim situates, and a copy of the recorded affidavit filed with the Bureau of Land Management before 30 December of the calendar year. Claim declared abandoned by operation of law for failure to comply.

AMALGAMATE - To capture gold with mercury that coats the gold and allows the gold particles to adhere to each other.

ASSESSMENT WORK - Annual performance of labor or improvements on the claim that meet certain minimum expenditure as required by law. Non-cumulative. Must perform each year.

BEDROCK - The base rock layer or stratum upon which the placer gravel rests.

BENCH - An elevated strip of stream side gravel remaining from earlier stream flow but now left high and dry due to a shift or down cutting of water in the stream channel. Occur along dry channels as well as live streams.

BLACK SAND(S) - Composed mainly of magnetite, a magnetic oxide of iron. May contain other heavy minerals such as ilmenite. One of the objectives in panning, the other being gold.

CLAIM - Acquisition of title to valuable mineral within certain fixed boundaries and predicated upon "discovery."

CLASSIFY; TO CLASSIFY - To screen gravel to a predetermined size.

COLOR - Gold. Usually applied to fine gold passing a 20 mesh screen and smaller but may include larger flakes and nuggets.

CONCENTRATE - Heavier metal(s) remaining after the lighter sands, and gravel, wash away by water as in panning, or air as in drywashing with a drywasher.

DISCOVERY - The finding of valuable mineral in place in such quantity and quality that a person of ordinary prudence would be justified in further expenditure of his labor and means, with a reasonable prospect of success in developing a valuable mine. A prerequisite for a claim.

DISCRIMINATION - The ability of a metal detector to identify ranges or detected targets.

DRYWASHER - A gold-concentrating device that processes placer gravel by means of intermittent puffs or continuous draft of air to allow the heavier gold to settle behind a series of cross bars (riffles).

DRYWASHING - The mechanical operation of a device commonly known as a drywasher to capture gold present in placer material.

FERROUS - Pertaining to iron and iron compounds.

FREE GOLD - Gold eroded from the host rock and that now occurs as particles in placers.

GOLD WHEEL - A motorized device using water sprayed over a rotating bowl containing spiral grooves, which separate gold from the concentrate.

HOT ROCK - A mineralized rock containing either a greater or lesser amount of magnetic iron than the search area (usually the ground) that produces a positive (but false) signal (response) in a metal detector.

LOCATION - Making a discovery of "mineral in place," erecting boundaries defining the extent of the claim and posting a notice (location notice) signed by the locator (claimant) in or on the "location" monument or post.

LOCATOR - One who makes the discovery of valuable mineral in place and posts a location document (notice) to acquire a claim.

LODE; LODE GOLD - Gold enclosed in the host, mother or country rock such as a vein or ledge of quartz.

METAL - As related to the operation of a metal detector, a ferrous or non-ferrous substance capable of conduction of electricity and producing a positive response from the detector.

NONFERROUS - Pertains to metals and minerals without any iron such as brass, silver, lead and aluminum.

NUGGET - Native placer gold larger than a grain in weight.

OVERBURDEN - The overlying sands, and gravel, usually devoid of placer gold.

PAN; TO PAN - Reducing the amount of gold-bearing sand and gravel by use of a miner's gold pan and water; to concentrate by use of a pan.

PAYDIRT; PAYSTREAK - The gold-bearing layer or layers of placer gravel, usually of limited extent.

PLACER - Pronounced "plasser." Generally applied to gold particles occurring in loose, unconsolidated material such as sand and gravel that may contain other heavy minerals.

POCKET - An accumulation of gold-bearing material in a concentrated area such as a depression in bedrock.

RIFFLE(S) - The cross bar(s) made of metal or wood that hinder(s) the migration of gold through a drywasher or sluice.

RIFFLE TRAY - As pertains to a drywasher, the rectangular frame with riffles.

SLUICE; SLUICE BOX - A U-shaped channel several feet in length made of metal or wood, containing riffles, through which directs flowing water and gold-bearing materials.

SUCTION DREDGE - A gold mining device using a suction created by the injection of water under high pressure into a nozzle connected by a flexible hose to a sluice box. In operation, the nozzle sucks up gold-bearing gravel and passes it through the hose to the sluice box where it flows over the riffles.

TAILINGS - The waste material that discharges from the drywasher or sluice in its operation.

WASH - A dry watercourse found in deserts that may have water flowing in it only after heavy rains.

***

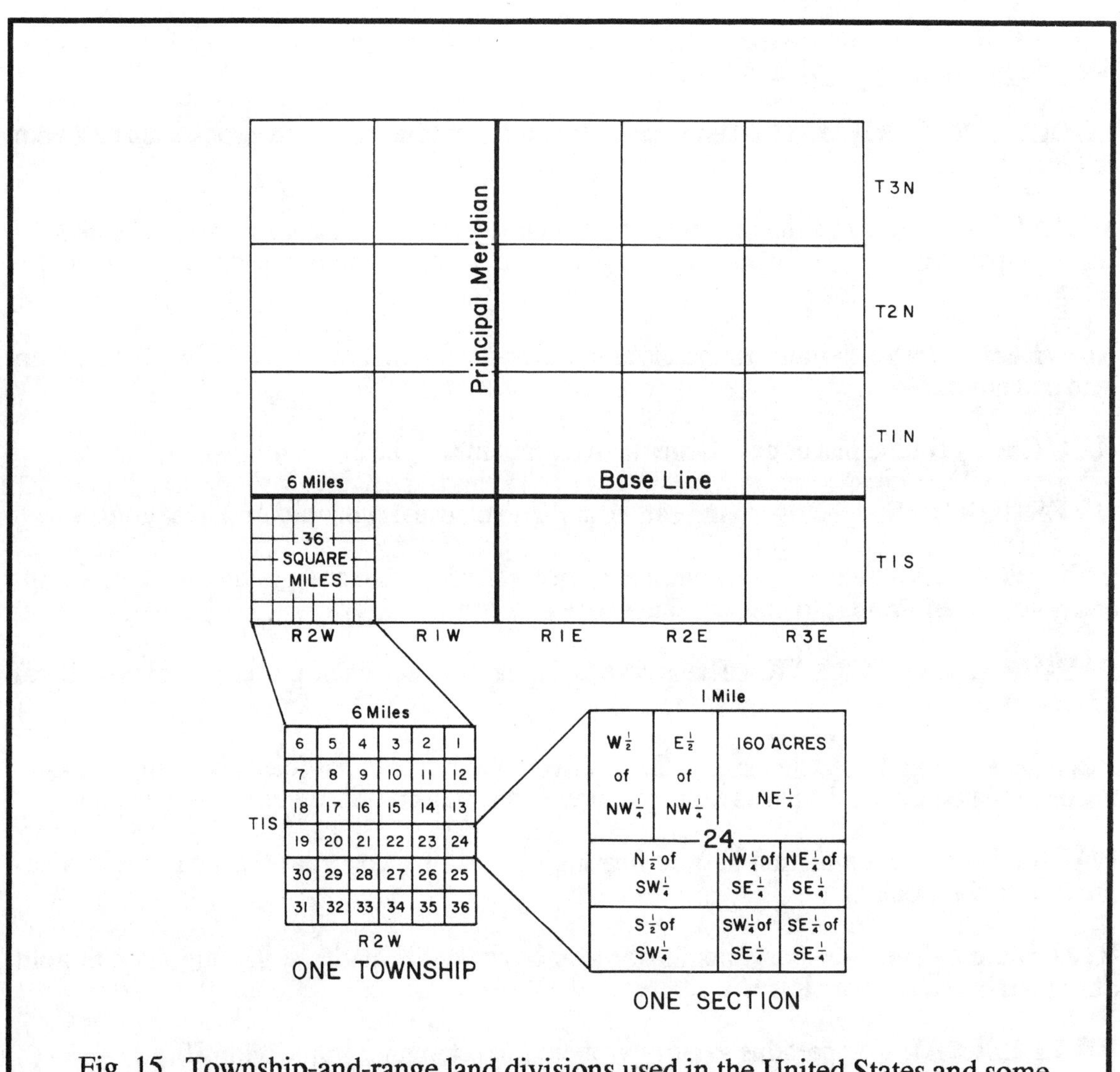

Fig. 15. Township-and-range land divisions used in the United States and some parts of Canada. From Zumberge and Rutford, 1983.